SMP 11-16

Teacher's guide to Book R1

CAMBRIDGE
UNIVERSITY PRESS

Published by the Press Syndicate of the University of Cambridge
The Pitt Building, Trumpington Street, Cambridge CB2 1RP
40 West 20th Street, New York, NY 10011–4211, USA
10 Stamford Road, Oakleigh, Melbourne 3166, Australia

First published 1986
Sixth printing 1993

Typesetting and diagrams by Marlborough Design
Printed in Great Britain at the University Press, Cambridge

ISBN 0 521 31473 9

Contents

The School Mathematics Project was founded in 1961 with the purpose of improving the teaching of mathematics in schools by the provision of new course materials. SMP authors are experienced teachers and each new venture is tested by schools in a draft version before publication.

Work on SMP 11–16 started in 1977 and the pilot version of the course has been used by some 50 schools, most of them comprehensive but including some selective schools, since 1980. The published version of the course started appearing in 1983.

Since its inception the SMP has always offered an 'after sales service' for teachers using its materials. If you have any comments on SMP 11–16 or would like advice on its use please write to
SMP Office
University of Southampton
Southampton
SO9 5NH

The following people have contributed to the planning and writing of the Y, B and R series of books.

Graham Ambridge   Phil Goodwin   Brian Hughes
Chris Belsom   Eric Gower   Spencer Instone
Neil Bibby   Harry Gurevitch   Sylvia Johnson
Michael Darby   Graham Hall   John Ling
Charles Dickinson   Joyce Harris   Alan Mace
David Fenton   Ray Harris   Paul Scruton
Tony Gardiner   Stephen Horner   Martyn Truman
                                           Richard Walker

Others too numerous to be mentioned individually have provided valuable advice and help. Among these are the mathematics staff and pupils of the pilot schools whose detailed comments on the draft version were essential in revising the course for publication.

The SMP 11–16 team is led by John Ling.

With unfailing care the bulk of the manuscripts were typed for the press by Muriel Hudson. The authors wish to give particular thanks to Sue Glover for her work in preparing the materials for publication.

The coloured books of SMP 11–16

The yellow, blue, red, green and amber series together make up the second part of the SMP 11–16 course, for pupils in Year 9 to Year 11 (ages 13+ to 16+).

(The booklet strands which form the first part of the course are fully described in the *Teacher's file for key stage 3*.)

The overall structure of the coloured books is set out in the diagram below.

The Y series is for the most able group of pupils (roughly speaking, the top 20% to 25% or so, although the proportion is likely to vary from school to school). The B and R series are for the 'middle' group (the next 35% to 40% or so). The G series is for lower ability pupils and the A series for those who find the G series too demanding.

The B series branches after *Book B2* to allow the more able of the pupils in the middle group to move ahead on to more demanding work in the R series. The mathematical content of the R series has much in common with that of *Books Y1, Y2* and *Y3*, but presentation and pace are often different.

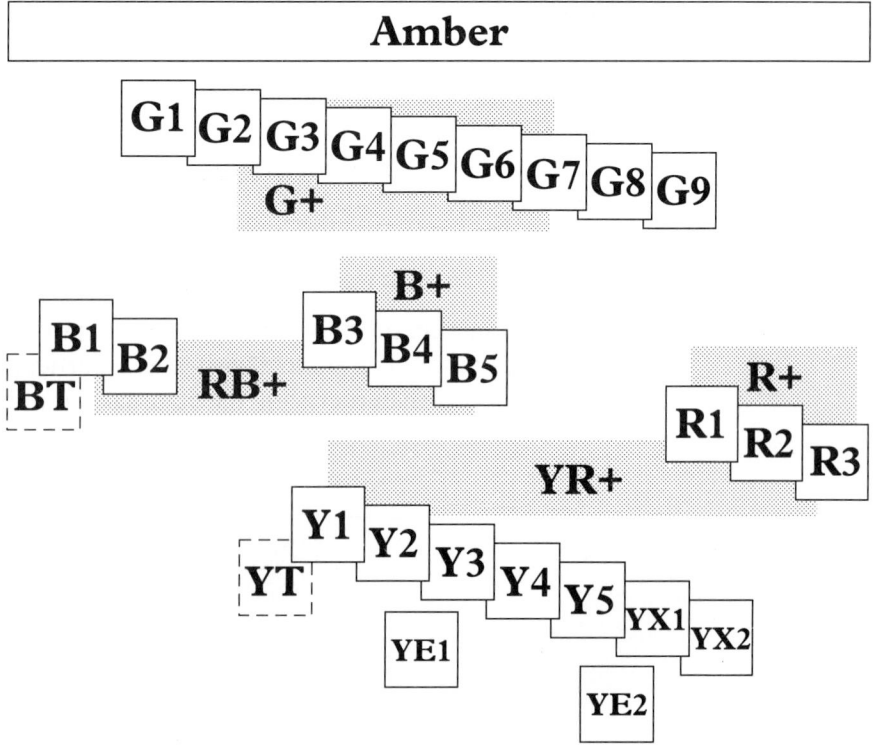

There are five supplementary books (Books $YR+$, $R+$, $RB+$, $B+$ and $G+$) which can be drawn on to ensure appropriate coverage of the national curriculum in key stages 3 and 4. Further information about these books is given in the *Teacher's file for key stage 3* and the *Teacher's file for key stage 4*.

The two YX books are 'extension' books written to stretch the most able pupils and cover content beyond that in *Book Y5*. *Books YT* and *BT* are transition books written for pupils who have not previously used the booklets.

Classroom organisation and teaching style

It is assumed that pupils will be grouped in sets according to ability in years 9, 10 and 11.

Although there is rather more exposition and explanation in the books than is found in many other textbooks, the books are not intended to be 'self-instructional'. Many important points arise in the course of doing the problems in the books, and these points will need to be brought out by the teacher in discussion with the whole class or with smaller groups or as appropriate. Teachers may find it possible from time to time to give particular chapters or sections of chapters to the class to work through on their own, which is no bad thing since the ability to pick up information from the printed page and to follow written explanations is an important one. Where this is done it will be necessary for the teacher carefully to 'go over' what has been done.

There are no 'chapter summaries'. The writers feel it is more valuable for classes to make their own summary notes. The ideal ultimately is for each pupil to make his or her own notes, but initially it may be better for the teacher to lead a discussion after each chapter of the main ideas before any notes are made.

Introduction to Book R1

Starting assumptions

Book R1 follows on from *Book B2* and is intended for pupils who have found *Books B1* and *B2* relatively straightforward and who are capable of moving on to more demanding work.

Mental and written arithmetic and the use of calculators

It is assumed throughout that unless there is an instruction to the contrary calculators will be used for all but the simplest calculations which can be done mentally.

We strongly recommend that teachers encourage mental calculation, and from time to time give short sets of questions to be answered mentally. We also suggest having occasional practice sessions on written arithmetic, but that the scope of these should not extend beyond addition, subtraction, multiplication by 2, 3, . . . 9 and division by 2, 3, . . . 9 of whole numbers and money.

Starred sections and questions

Occasional sections and questions are starred to indicate that they are of greater difficulty and can be left out by slower pupils using the book.

Equipment needed

Certain standard items of equipment are needed frequently and no special attention is drawn to them in the books. These include rulers, angle measurers (recommended for angle measurement; see below), compasses, scissors and 2 mm graph paper.

In other cases, equipment needed (such as tracing paper) is referred to in the book. Worksheets are needed occasionally. Masters for these are available separately (see below). Seven worksheets are needed for *Book R1*, numbered R1–1 to R1–7.

Pupils working from the R series are assumed to have the use of a scientific calculator.

Ordering equipment

The following items required for *Book R1* are published by Cambridge University Press. You should order them through your usual school book supplier.

Worksheet masters for the B and R series   ISBN 0 521 33625 2
Angle measures (pack of 5)   ISBN 0 521 25435 3
Pie chart scales (pack of 10)   ISBN 0 521 26366 2

When ordering, remember to state the ISBN, the series title (SMP 11–16), the name of the item, the publisher and the number of **packs** you want. (So, for example, if you want 35 angle measurers, write your order as '7 packs of 5'.)

Notes and answers for Book R1

1 Solids

Most of the work in this chapter is on cubes and cuboids. The start of section D revises work on prisms met earlier in the booklet *Volume* (or in *Book BT*). Some other kinds of solid are introduced in section E.

A Nets of a cube and a cuboid

A1 A, C and D are nets of a cube.

A2 Here is one possible net:

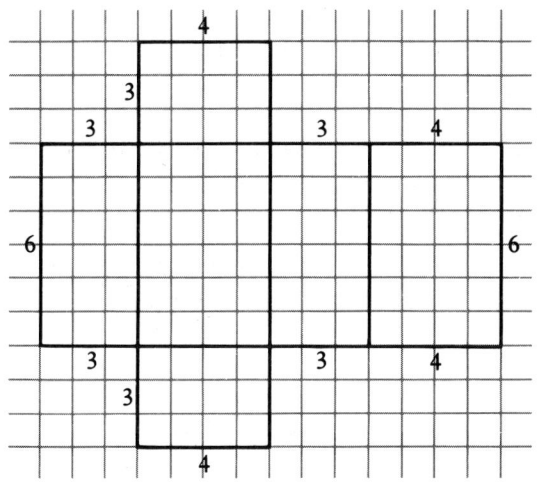

A3 A, D and E are nets of a cuboid.

A4 A sketch like the one shown on page 3 of the pupils' book

B The volume of a cuboid

B1 (a) $30\,cm^3$ (b) $120\,cm^3$
(c) $1000\,cm^3$ (d) $64\,cm^3$
(e) $30\,cm^3$ (f) $76 \cdot 44\,cm^3$
Discuss the accuracy of answer (f).

B2 $78\,m^3$

B3 $15 \cdot 3\,m^3$

B4 (a) 600 ml (b) 650 ml
(c) 830 ml (d) 4830 ml

B5 (a) 170 ml (b) 540 ml
(c) 360 ml (d) 361 ml

B6 (a) 0·3 litre (b) 0·35 litre
(c) 0·94 litre (d) 0·943 litre

B7 The three bottles together hold 1·704 litres.
The two bottles together hold 1·7 litres.
So the three bottles hold very slightly more, 4 ml more.

B8 (a) $12\,960\,cm^3$ (b) 13 litres

B9 (a) $2 \cdot 4\,m^3$ (b) 2400 litres

B10 (a) $537 \cdot 6\,m^3$ (b) 537 600 litres
(c) 672 minutes (d) 11 hours

B11 171 litres

C Packing cubes

C1 $50\,cm^2$

C2 (a) Here is one possible net:

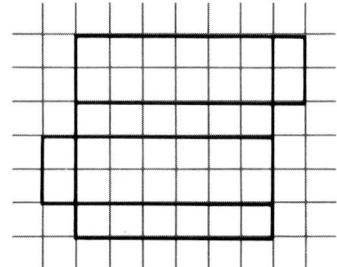

C2 continued

(b) $40\,cm^2$

(c) The second box needs **less** card than the first one.

C3 Here is one possible net:

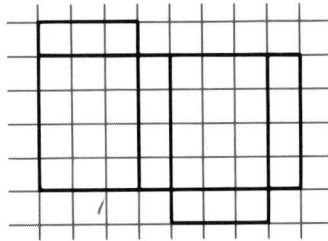

The area for any correct net will be $38\,cm^2$, so it uses **less** cardboard.

C4 A cuboid $2\,cm$ by $2\,cm$ by $3\,cm$ uses only $32\,cm^2$ of cardboard.

D Prisms

D1 A and C are prisms.

D2 (a) $66\,cm^3$ (b) $32\,cm^3$

D3 (a) $16\,cm^2$ (b) $160\,cm^3$

D4 $260\,cm^3$

D5 $210\,cm^3$

★D6 $3 \cdot 3\,m$ deep

★D7 About $0 \cdot 55\,m$ thick

★D8 $40\,cm$ long

★D9 $0 \cdot 017\,cm$ thick

★D10 $4000\,cm^2$

★D11 $500\,cm^2$

★D12 $40\,m^2$ or $400\,000\,cm^2$

★D13 $1000\,cm^2$

E Other solids

E1 (a), (b), (d) and (f) are polyhedra.

E2 (a) Square-based pyramid
(b) Triangular prism

E3 Pupil will have made a cone.

E4 A 1, B 3, C 2

E5 (a)

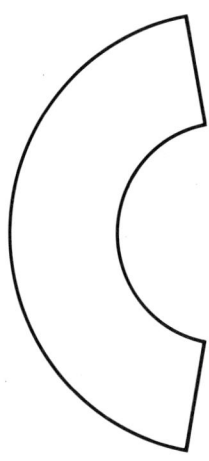

(b) Pupil will have made a 'lampshade'.

E6 D is half full.

---

2 Approximation

This chapter introduces significant figures, and the method of estimating by rounding off numbers to one significant figure.

A Rounding off: a reminder

A1 (a) $90\,000$ (b) $34\,000$
(c) $35\,000$ (d) $279\,000$
(e) $152\,000$ (f) $63\,000$

A2 (a) 600 (b) 900 (c) 1200
(d) 3500 (e) 8900 (f) 4100
(g) $16\,700$ (h) $35\,900$ (i) $13\,000$
(j) $850\,100$

A3 (a) 0·9 (b) 0·3 (c) 0·6 (d) 7·4
(e) 5·1 (f) 7·8 (g) 1·1 (h) 92·4
(i) 48·1 (j) 51·3

A4 (a) 6·27 (b) 69 300 (c) 45 300
(d) 0·1 (e) 5·7 km (f) 5·90

A5 (a) 4·13 (b) 128·46 (c) 1·01
(d) 2·06 (e) 5·13 (f) 2232·86

B Significant figures in numbers greater than 1

B1 (a) 5000 (b) 8000 (c) 80 000
(d) 50 000 (e) 400 (f) 500
(g) 20 000 (h) 100 000 (i) 10 000
(j) 10 000

B2 (a) 3800 (b) 7300 (c) 41 000
(d) 71 000 (e) 150 000

B3 (a) 458 000 (b) 26 600 (c) 44 500
(d) 30 100 (e) 70 200 (f) 109 000
(g) 86 400 (h) 1 440 000
(i) 784 000 (j) 100 000

C Significant figures in decimals

C1 (a) 0·041 37 (b) 0·006 84 (c) 3·0082

(d) 17·6039 (e) 0·050 23 (f) 1·009 37

(g) 0·008 71 (h) 0·930 26

C2 (a) 0·07 (b) 80 (c) 0·008
(d) 0·0005

C3 (a) 0·29 (b) 37 (c) 0·082
(d) 0·0011 (e) 0·20 (f) 160
(g) 0·075 (h) 0·061

C4 (a) 3·7 (b) 0·37 (c) 4 (d) 0·095
(e) 1·1 (f) 0·0069

C5 (a) 27·7 (b) 145 (c) 6·01
(d) 0·003 26

D Multiplication patterns

D1 (a) 60 (b) 600 (c) 6000 (d) 60 000

(e) 2000 (f) 18 000 (g) 80 000
(h) 210 000

D2 (a) 0·8 (b) 0·08 (c) 0·008 (d) 0·3

D3 (a) 0·24 (b) 0·08 (c) 0·015
(d) 0·024 (e) 0·06 (f) 0·4

D4 (a) 15 000 (b) 0·008 (c) 350 000
(d) 0·004

D5 (a) 2·4 (b) 14 (c) 4
(d) 180 (e) 150 (f) 20

D6 (a) 3·2 (b) 0·003 (c) 0·06

E Rough answers

E1 (a) 6, 6·955 (b) 12, 12·2544
(c) 32, 30·2187 (d) 150, 145·9778
(e) 240, 243·162 (f) 1500, 1381·915

E2 (a) 0·12, 0·1134 (b) 0·15, 0·1504
(c) 5·4, 5·521 25 (d) 0·8, 0·835 36
(e) 16, 16·400 24 (f) 0·06, 0·0646

E3 (a) £140 (b) £126·73

E4 A typical reason might be
because 0·6 × 700 = 420.

E5, E6, E7 require discussion of pupil's
own methods and answers. Some
typical answers might be:

E5 (a) About 6·5 m, 4·0 m
(b) About 25 m² (c) About £200

E6 (a) About £100 (b) About £100
(c) About £70 (d) About £30

E7 (a) About 10 m² (b) About 4 m²
(c) About 5 m² (d) About £200

F Significant figures and decimal places

F1 (a) 470 (b) 472·62

F2 (a) 28·6 (b) 28·563

F3 (a) 8·1 (b) 0·07 (c) 0·0849
(d) 101 (e) 8·039 (f) 625

3 Investigations (1)

Each of these investigations involves generalisation.

1 Rectangles

(a) 10   (b) 15   (c) 21

(d) The increase goes up by 1 each
time: $6 + \mathbf{4} = 10$, $10 + \mathbf{5} = 15$,
$15 + \mathbf{6} = 21$, etc.
For the diagram with 12 bricks,
the number of rectangles will be
$21 + 7 + 8 + 9 + 10 + 11 + 12 = 78$.

(e) Think of going from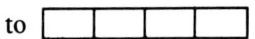

to

by adding an extra 'brick'.
What rectangles are there in the
second diagram which have not
already been counted in the first
one?

An alternative approach is to count
as in the example below:

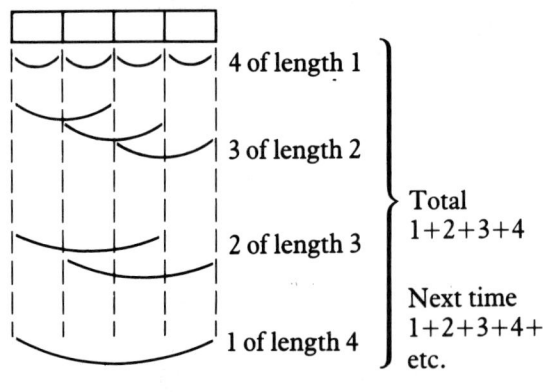

4 of length 1

3 of length 2

2 of length 3

1 of length 4

Total
$1+2+3+4$

Next time
$1+2+3+4+5$
etc.

(f) (i) 18   (ii) 30   (iii) 36

In (i) you can think of the diagram as a
combination of three 3-brick strips:

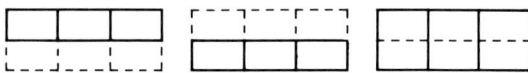

In (ii) there are three 4-brick strips.

In (iii) there are six 3-brick strips.

A possible sequel to this investigation is the
problem 'How many squares are there on a
chessboard?' (See also *Stretcher* number 10 in
Investigations and stretchers.)

2 Bridges

Some way of recording the routes is useful.
Trying to do it on the diagram itself can lead
to confusion. Labelling the bridges helps.
Suppose the bridges on each river are
numbered: 1, 2, 3 on the first river; 1, 2 on
on the second; 1, 2 on the third. The route
shown can be written as $3 - 1 - 1$.

The number of routes on the given map is
12. This number is $3 \times 2 \times 2$, which
illustrates the general rule whereby we
multiply together the number of bridges on
the separate rivers to get the number of
routes.

4 Angles and bearings

Bearings were first introduced in the booklets *Turning* and *Bearings and journeys*. This chapter contains work on bearings and angles of elevation.

All answers dependent on measuring angles are liable to vary by at least ± 1°. These answers should not be assumed to be the best possible answers, just approximate answers, except where it is indicated that they have been calculated.

A Bearings from a point

A1 *a* 34°, *b* 139°, *c* 195°, *d* 243°, *e* 306°

A2 *a* 022°, *b* 055°, *c* 100°, *d* 217°, *e* 253°
f 308°

A3 *Swift* 197°, *Intruder* 252°, *Marigold* 290°

A4 *Barnacle* 2·8 km, *Swift* 3·9 km,
Intruder 5·7 km, *Marigold* 5·4 km

A5 (a), (b)

B Fixing a position

B1

Distance from P, 6·5 km (calculated)
Distance from Q, 4·4 km

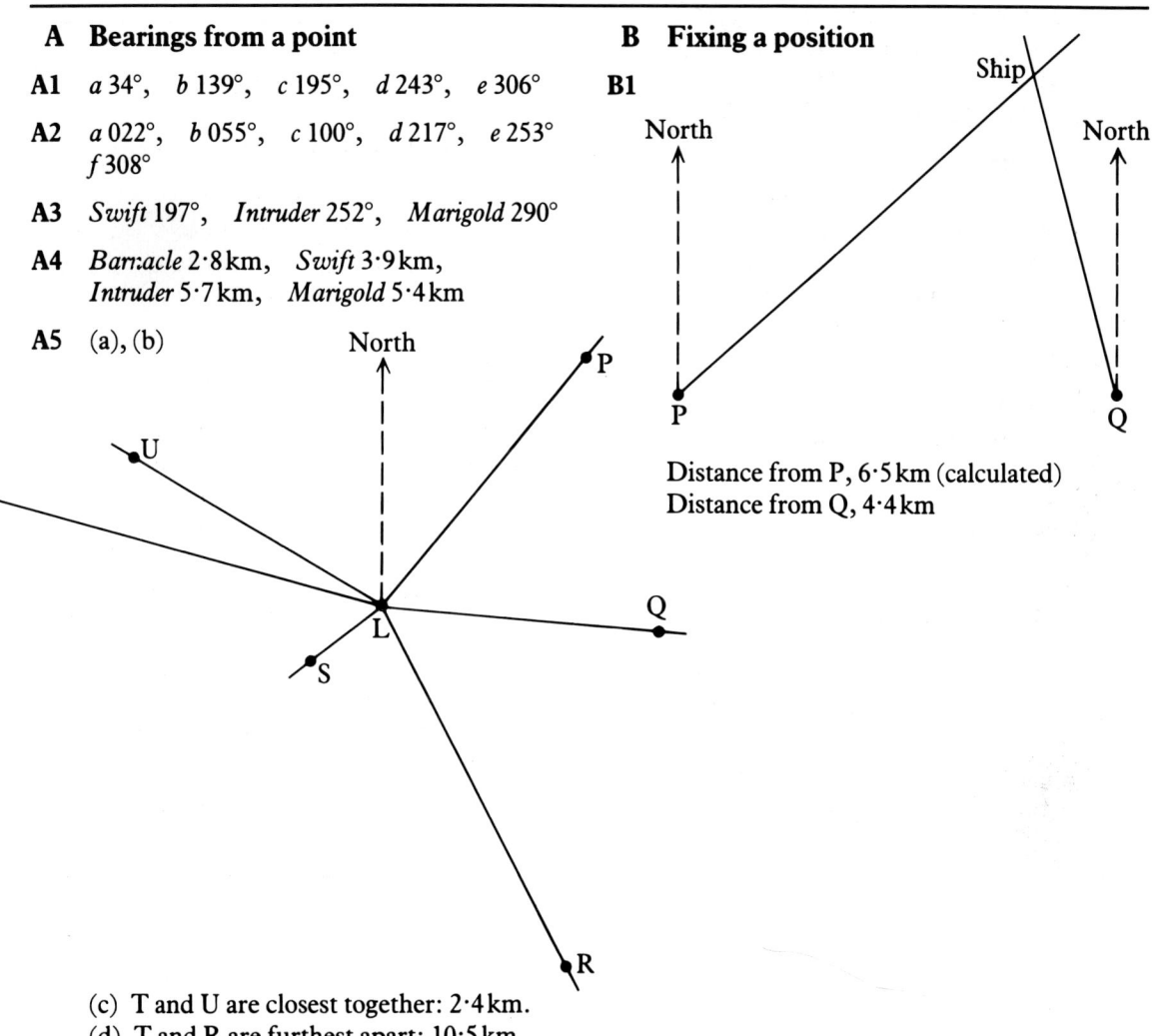

(c) T and U are closest together: 2·4 km.
(d) T and R are furthest apart: 10·5 km.

B2

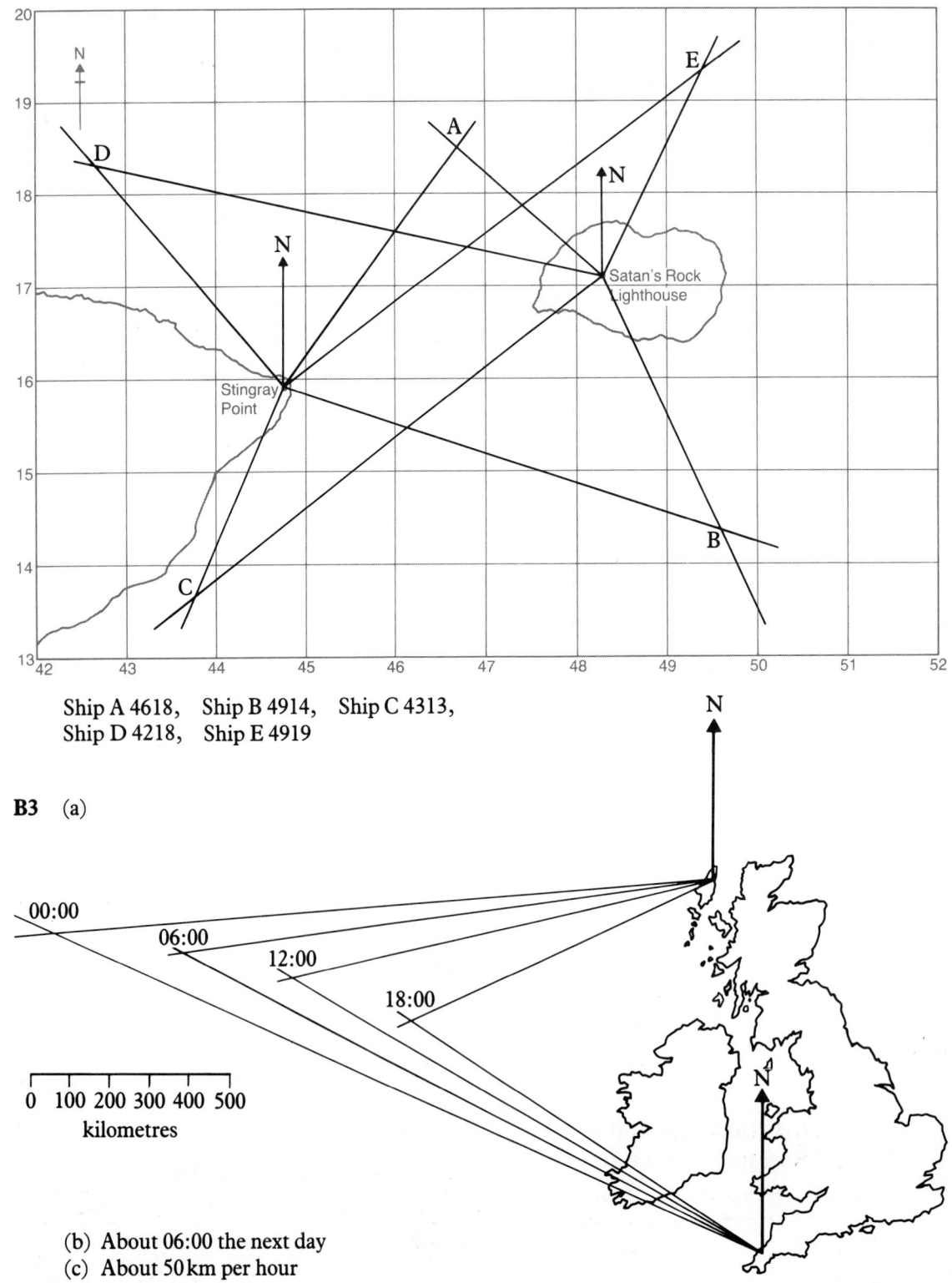

Ship A 4618,   Ship B 4914,   Ship C 4313,
Ship D 4218,   Ship E 4919

B3   (a)

(b) About 06:00 the next day
(c) About 50 km per hour

The following answers have been calculated to 3 s.f.

C1 6·25m   **C2** 64·0m   **C3** 15·6m   **C4** 45·6m   **C5** 31·1m

---

5 Looking at data

For relatively small sets of data, marking each item on a scale can give a useful picture of the distribution.

On page 27, the groupings 1·5–2·0, 2·0–2·5, etc. are used, so that the frequency chart can have an ordinary continuous scale. This involves a decision as to what to do about weights of exactly 2·0kg, etc. An alternative is not to use a scale at all but simply to label the bars '2·0 to 2·4', '2·5 to 2·9', etc.; a more sophisticated method is to use intervals of 1·45–1·95, 1·95–2·45, etc. so that none of the weights is on a boundary, but to do this would cloud the issue at this level.

A Frequency charts

A1 (a) 2 (b) 7 (c) 9 (d) 15 (e) 11
    (f) 5 (g) 1

A2 (a) 47 (b) 37 (c) 31
    (d) 3·5–4·0kg is the modal group.

A3 (a) Type A tended to produce heavier tomatoes.
    (b) 60–80 grams is the modal group for A. 20–40 grams is the modal group for B.
    (c) 41 (d) 10 (e) 24% (to 2 s.f.)
    (f) 53% (to 2 s.f.)

B The median of a set of measurements

B1 (a) Number 6 (b) 52kg

B2 (a) Number 12 (b) 5·2m

B3 (a) Numbers 8 and 9 (b) 168cm

B4 (a) 20 (b) Numbers 10 and 11
    (c) 164·5cm

B5 For n items, the numbers of the middle pair are $\frac{1}{2}n$ and $\frac{1}{2}n + 1$.

B6 For n items, number of middle one is $\frac{1}{2}(n+1)$. [Note: pupils will write these rules in their own way, not necessarily algebraically.]

B7 (a) (i) Number 8 (ii) 52kg
    (b) (i) Numbers 10 and 11 (ii) 55·5kg

C Summarising data

C1

	Smallest	Median	Largest	Range
(a)	64	82	100	36
(b)	22	41	57	35
(c)	3·4	5·35	6·9	3·5

C2 (a)

Soil A

Soil B

(b)

	Shortest	Median	Longest	Range
Soil A	31	46·5	73	42
Soil B	37	61	74	37

(all measurements in mm)

15

(c)

Soil A

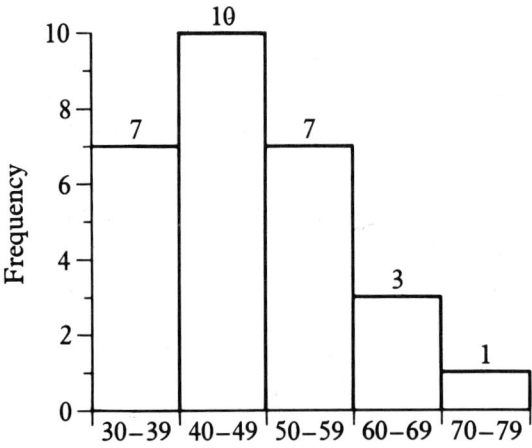

Length in mm

Soil B

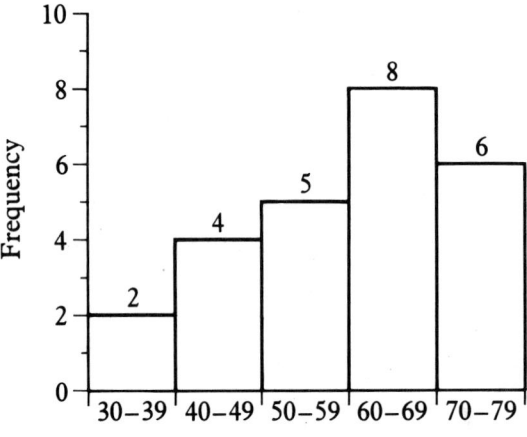

Length in mm

(d) **Brief report**

As a group, the worms in soil B are longer than the worms in soil A.

The worms in soil A are more widely spread out in length.

★D   Stem-and-leaf tables

★D1   (a)

4	5	8			
5	3	3	5	6	
6	0	1	4	4	7	8
7	4	6			
8	3				

(b) 60–69 is the modal group.
(c) 61 is the median mark.
(d) 38 is the range.

★D2   (a)

1·	6	8				
2·	3	5	5	6	7	8	8
3·	0	0	3	5	6	9
4·	0	1	1	2	3	

(b) 3·0 kg

★D3   (a) Paper 2 was harder than paper 1 because fewer candidates got high marks and more got low marks.
(b) 66 is the median mark for paper 1.
50 is the median mark for paper 2.
(c) 72 is the range for paper 1.
74 is the range for paper 2.

Dissection puzzles

It was not felt necessary to include an answer here!

6 Squares and square roots

This chapter introduces square roots. The decimal search method can easily be extended to finding approximate cube roots (with the advantage that there is usually no cube root key on the calculator to do the work more easily!).

The chapter introduces Pythagoras' rule, which is then used to calculate the longest side of a right-angled triangle. Calculation of one of the shorter sides comes later, in chapter 14.

A   Square roots

A1   (a) 36   (b) 49   (c) 64   (d) 81   (e) 100

A2   (a) 529   (b) 3249   (c) 164025
(d) 698896

A3   (a) 2209   (b) 105625   (c) 250000

A4   (a) 3   (b) 2   (c) 6

A5   (a) 8×8 is $\textcircled{64}$, so 8 is the square root of $\textcircled{64}$.
(b) The square of 10 is $\textcircled{100}$, so 10 is the square root of $\textcircled{100}$.
(c) 7×7 is $\textcircled{49}$, so $\textcircled{49}$ is the square of $\textcircled{7}$, and $\textcircled{7}$ is the square root of $\textcircled{49}$.

B   The squares graph

B1   (a) 0·2   (b) 2

B2   (a) About 6·3   (b) Yes, $6 \cdot 3^2 = 39 \cdot 69$

B3   About 5·5, $5 \cdot 5^2 = 30 \cdot 25$

B4   (a)

Number	0 . . .	8	9	10
Square of number	0 . . .	64	81	100

The graph will, of course, be an extended version of that on page 39.

(b) About 8·7   (c) $8 \cdot 7^2 = 75 \cdot 69$

C   Decimal search

C1   The final entries will be

Number . . .	4·472	4·473	. . .
Square . . .	19·998 784	20·007 729	. . .

C2   Entries in the table will vary.
$\sqrt{110} = 10 \cdot 488\,088$ (to 6 d.p.)

C3   Entries will vary.
$\sqrt{150} = 12 \cdot 247\,449$ (to 6 d.p.)

D   Rounding off square roots

D1   (a) 3·87 cm   (b) 7·75 cm

D2   (a) 1·60 s   (b) 0·80 s
(c) 1·34 s   (d) 1·71 s

D3   (a) 21·1 km   (b) 33·1 km

E   Pythagoras' rule

E1   (a) 8·94 cm   (b) 5·83 cm
(c) 10·00 cm

E2   (a) 5·83 m
(b) 5·8 m is a good answer for a scale drawing.

E3   10·5 cm

E4   6·36 cm (to 2 d.p.)   **E6**   16·0 m

E5   5·6 m   **E7**   427 m

Review 1

1 Solids

1.1 B and C are nets of cuboids.

1.2 $1 \cdot 1 \, m^3$

1.3 (a) The four bottles contain 2·920 litres.
The two bottles contain 2·9 litres.
So the four bottles hold more.
(b) 20 millilitres

1.4 (a) $77 \, cm^3$ (b) $156 \, cm^3$ (c) $90 \, cm^3$

1.5 $144 \, cm^3$

★1.6 $800 \, cm^2$ **★1.7** $56 \, cm$

2 Approximation

2.1 (a) 47 000 (b) 4·7
(c) 583 300 (d) 0·09
(e) 6·8 kg (f) 7·0

2.2 (a) 500 (b) 8000
(c) 20 000 (d) 900 000
(e) 30 000

2.3 (a) 5100 (b) 3500
(c) 12 000 (d) 110 000
(e) 89 000

2.4 (a) 0·0004 (b) 0·07
(c) 0·0003 (d) 0·1

2.5 (a) 0·044 (b) 0·000 67
(c) 0·000 81 (d) 4·7

2.6 (a) 0·0361 (b) 2·84
(c) 42·9 (d) 6·80

2.7 (a) 2000 (b) 21 000
(c) 40 000 (d) 720 000

2.8 (a) 0·35 (b) 0·1
(c) 0·024 (d) 0·04

2.9 (a) 15 (b) 4·2
(c) 6 (d) 2

2.10 (a) 320 (b) 2400 (c) 240
(d) 2 (e) 10 (f) 0·63

2.11 (a) $42 \, m^3$ (b) $40 \cdot 4 \, m^3$

3 Angles and bearings

3.1

	Distance	Bearing
Grange Farm	9·4 km	035°
Ransome's Farm	16·2 km	075°
Turbey Farm	7·8 km	101°
Packers Farm	17·7 km	101°
Strockley Farm	8·9 km	228°
Red Beck Farm	10·5 km	315°

3.2

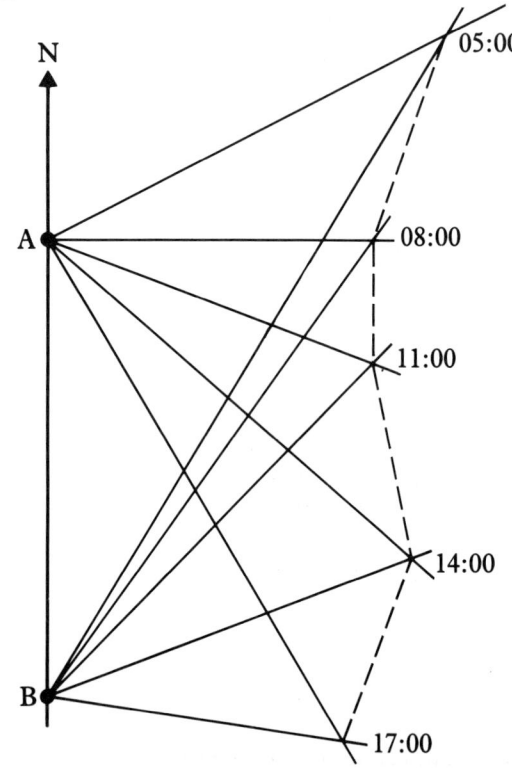

3.3 25·6 m (to 3 s.f.) by calculation

4 Investigations (1)

The pattern of folds develops in a very predictable way which pupils can be asked to try to explain.

```
after 1 fold                        D
      2              U              D              D
      3       U      U      D       D       U      D       D
      4    U  U  D   U  U   D   D   D   U   U   D   D   U   D   D
      5  U U D U U D D U U U D D U D D D U U D U U D D D U U D D U D D
```

There is an opportunity to extend the task so that a pupil may be asked
what the pattern would be after, say, 10 folds (if that many were possible);
or what if the strip were folded in thirds each time?

5 Looking at data

5.1 (a) 48·3 kg (b) 56·7 kg
 (c) 48·3, 49·8, 50·2, 51·7, 53·8, 54·4, 56·7
 The median is 51·7 kg.
 (d) 8·4 kg

5.2 (a) 15 kg (b) 51 kg (c) 50 kg

5.3 (a)

 (b) (i) 50 cm (ii) 89 cm
 (iii) 66 cm (iv) 39 cm

5.4 (a) The boys have greater armspans on the whole. This can easily be seen by superimposing the two frequency charts.
 (b) (i) 170−180 cm
 (ii) 150−160 cm
 (c) 65% (d) 7·5%

6 Squares and square roots

6.1 (a) 25 (b) 49 (c) 100

6.2 (a) 4 (b) 1 (c) 8

6.3 (a) ③ is the square root of 9.
 (b) ⑧① is the square of 9.
 (c) 400 is the square of ②⓪.
 (d) 50 is the square root of ②⑤⓪⓪.

6.4 38, 40 and 45 have square roots between 6 and 7.

6.5 The time for the deeper well is less than twice the time for the less deep. (In fact the relationship between the times is $t_2 = \sqrt{2}\,t_1$.)

6.6 (a) 6·40 (b) 6·32
 (c) 7·72 (d) 12·04

6.7 32 m

M Miscellaneous

M1 (a) 1600 (b) 12
 (c) 15 000 (d) 0·04
 (e) 21 (f) 120 000
 (g) 176 (h) 43·8

M2 (a) A quarter (b) 16 tiles
 (c) 320 tiles (d) 27 boxes
 (e) £108

M3 $a = 97°$, $b = 83°$,
 $c = 146°$, $d = 35°$,
 $e = 82°$, $f = 63°$

M4 (a) $p = 37$ (b) $s = 7$

M5 (a) 151 (b) 15·2% (to 3 s.f.)
 (c) 66·2% (to 3 s.f.)

7 The language of algebra

This chapter revises and extends the scope of basic algebraic notation.

A A review of some shorthand

A1 (a) 8 (b) 15 (c) 9 (d) 2
(e) 8 (f) 6 (g) 2 (h) 5
(i) 9 (j) 6

A2 (a) 20 (b) 24 (c) 30
(d) 90 (e) 75

A3 (a) 240 (b) 200 (c) 400
(d) 200 (e) 4000

B Brackets

B1 (a) 16 (b) 13 (c) 19 (d) 27
(e) 4 (f) 1 (g) 40 (h) 14

B2 (a) $3 + (5 \times 4) = 23$
(b) $10 - (2 \times 3) = 4$
(c) $(10 - 2) \times 3 = 24$
(d) $8 \div (2 \times 2) = 2$
(e) $40 \div (2 + 3) = 8$
(f) $(40 \div 2) + 3 = 23$

B3 (a) $(5 \times 4) - (2 \times 3) = 14$
(b) $5 \times (4 - 2) \times 3 = 30$
(c) $6 + (2 \times 3) + 4 = 16$
(d) $(6 + 2) \times (3 + 4) = 56$

B4 (a) 23 (b) 17 (c) 22 (d) 8
(e) 32 (f) 38 (g) 2 (h) 18

B5 (a) 7 (b) 23 (c) 9 (d) 44
(e) 47 (f) 28 (g) 10 (h) ⁻2

B6 (a) 40 m/s
(b) 0 m/s; the bullet begins to fall back to earth.

B7 (a) 20 (b) 35

B8 (a) 100 (b) 63

B9 (a) 30 (b) 36 (c) 20 (d) 10
(e) 6 (f) 14 (g) 40 (h) 0

C Squaring

C1 (a) 13 (b) 6 (c) 32 (d) 5

C2 (a) 51 (b) 17 (c) 8 (d) 2

(e) 13 (f) 12 (g) 4 (h) 16

C3 (a) 13 (b) 6 (c) 28
(d) 31 (e) 10

C4 54

C5 $3a^2 = 3 \times a^2 = 3 \times 5^2 = 75$

C6 18

C7 (a) 18 (b) 32 (c) 75 (d) 27
(e) 50 (f) 160

C8 (a) $s = c - 5t^2$
$= 80 - (5 \times 3^2)$
$= 80 - 45$
$= 35$
(b) (i) 75 (ii) 60 (iii) 0

C9 (a) £8 (b) £13·50 (c) £20
(d) £27·50 (e) £36 (f) £80

C10 (a) 18 (b) 36 (c) 36 (d) 12 (e) 18
(f) 9 (g) 36 (h) 12 (i) 18 (j) 36

C11 (a) 9 (b) 6 (c) 20 (d) 0
(e) 30 (f) 12 (g) 36 (h) 19

C12 (a) $(a + b)c = 21$ (b) $(ab)^2 = 100$
(c) $(a + b)^2 = 49$ (d) $2(ac)^2 = 72$
(e) $(\frac{1}{2}a)^2 = 1$ (f) $b(c + a) = 25$

C13 $2(c - b)$

C14 $2c - b$

C15 $c^2 + 3a$ or $3a + c^2$

C16 $3(b^2 + a)$ or $3(a + b^2)$

C17 (a) A number b is added to 3 times a.
(b) A number a is added to a number b, and the result multiplied by 3.

D Chips

D1 (a) $3a + b$ (b) 17

D2 (a) $a^2 + b^2$ (b) $2(4a + b)$ (c) $(a^2 + 4b)^2$
(d) $(2a)^2 + 2b^2$ (e) $(ab)^2$ (f) $b(a + 2)$

20

D3 (a)

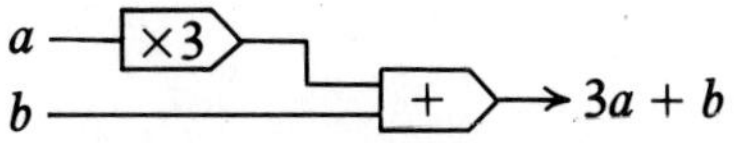

$a \xrightarrow{\times 3}$
b → $+$ → $3a + b$

(b)

a
$b \xrightarrow{+3}$ → $\times$ → $(b + 3)a$

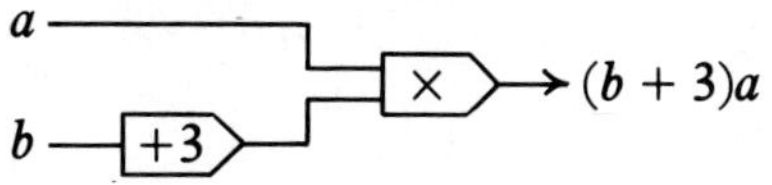

(c)

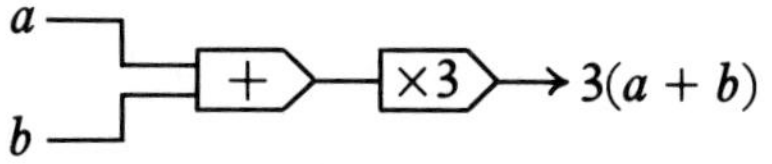

a
b → $+$ → $\times 3$ → $3(a + b)$

(d)

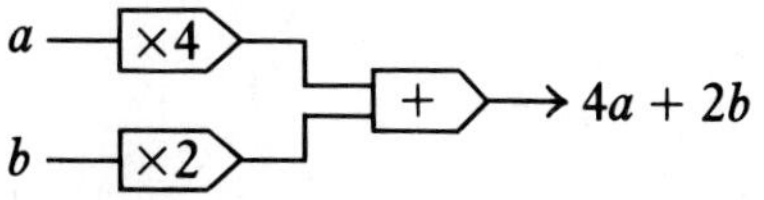

$a \xrightarrow{\times 4}$
$b \xrightarrow{\times 2}$ → $+$ → $4a + 2b$

(e)

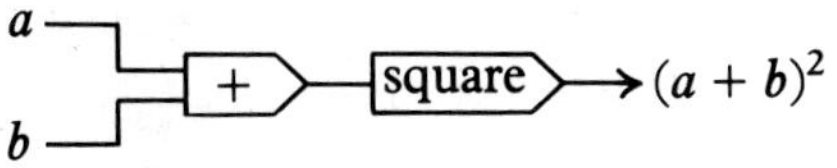

a
b → $+$ → square → $(a + b)^2$

(f)

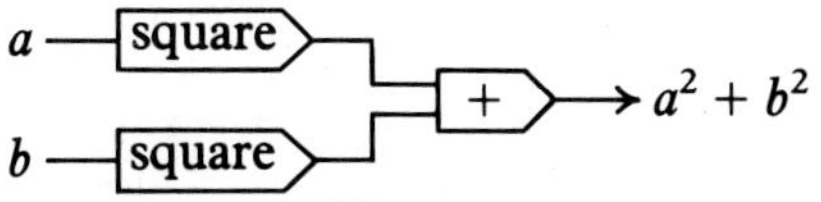

$a \xrightarrow{\text{square}}$
$b \xrightarrow{\text{square}}$ → $+$ → $a^2 + b^2$

(g)

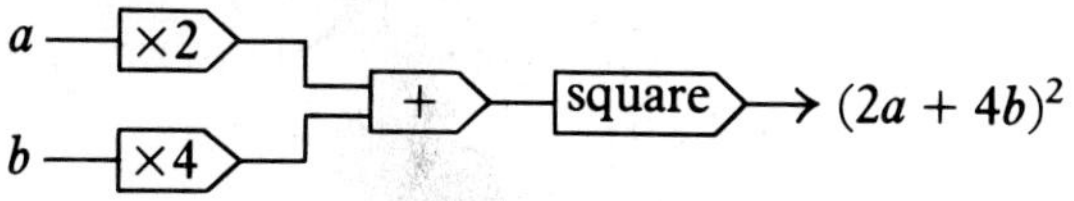

$a \xrightarrow{\times 2}$
$b \xrightarrow{\times 4}$ → $+$ → square → $(2a + 4b)^2$

(h)

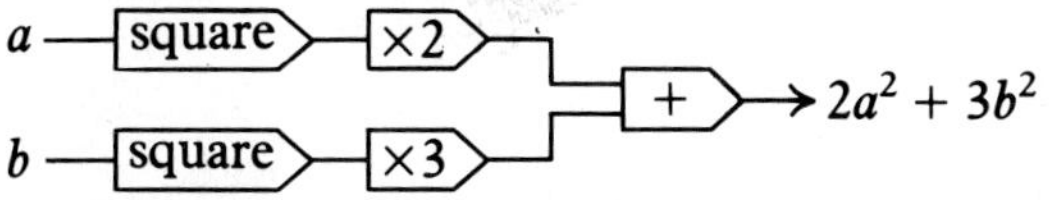

$a \xrightarrow{\text{square}} \xrightarrow{\times 2}$
$b \xrightarrow{\text{square}} \xrightarrow{\times 3}$ → $+$ → $2a^2 + 3b^2$

(i)

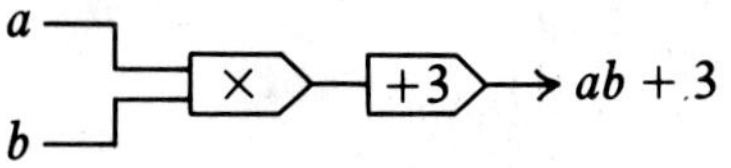

a
b → $\times$ → $+3$ → $ab + 3$

(j)

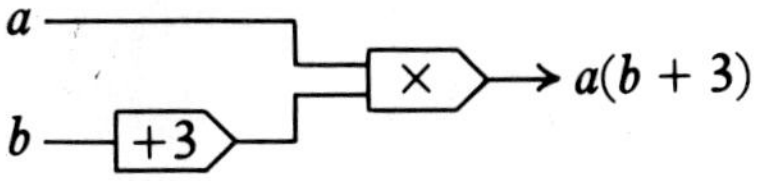

a
$b \xrightarrow{+3}$ → $\times$ → $a(b + 3)$

(k)

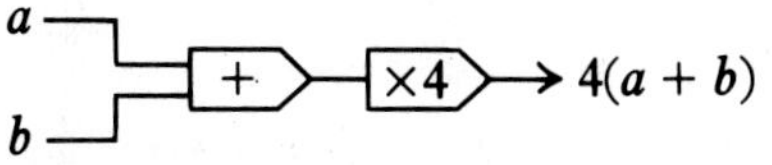

a
b → $+$ → $\times 4$ → $4(a + b)$

(l)

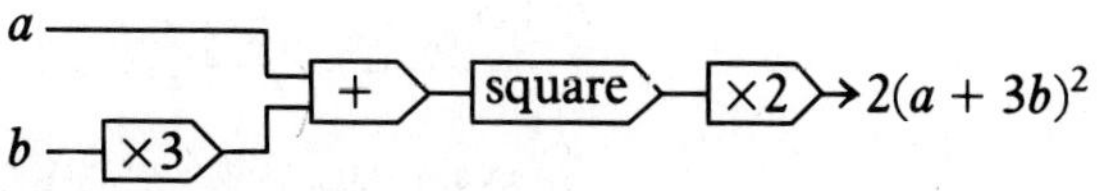

a
$b \xrightarrow{\times 3}$ → $+$ → square → $\times 2$ → $2(a + 3b)^2$

D4 (a) $c(3a + b)$ (b) $3ab + c^2$
(c) $3a + bc$ (d) $c(a + b)^2$

D5 Pupils may have alternative versions
for (b), (d) and (f).

(a)

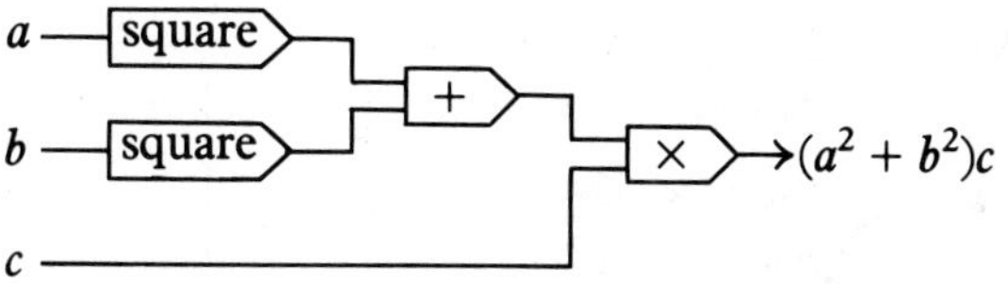

$a \xrightarrow{\text{square}}$
$b \xrightarrow{\text{square}}$ → $+$ → $\times$ → $(a^2 + b^2)c$
c

(b)

$a \xrightarrow{\times 2}$
b → $\times$ → $+$ → $2ab + 3c^2$
$c \xrightarrow{\text{square}} \xrightarrow{\times 3}$

(c)

a
b → $\times$ → $a(b + c)$
c → $+$

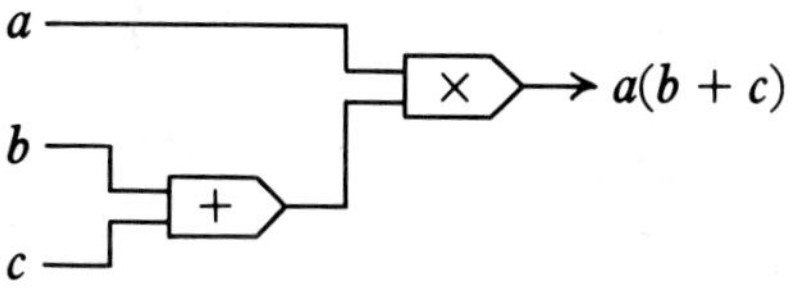

(d)

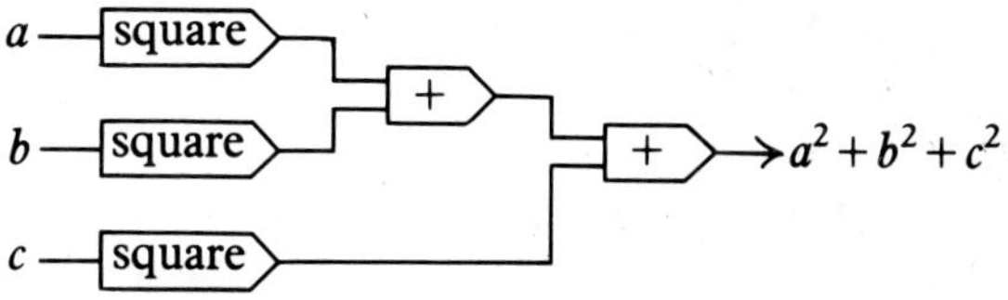

$a \xrightarrow{\text{square}}$
$b \xrightarrow{\text{square}}$ → $+$ → $+$ → $a^2 + b^2 + c^2$
$c \xrightarrow{\text{square}}$

(e)

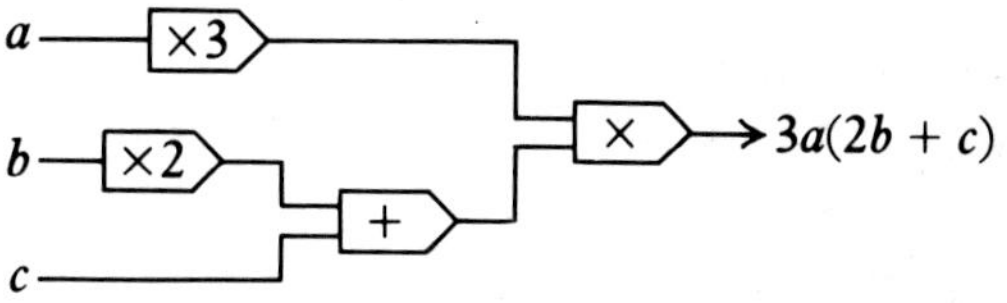

$a \xrightarrow{\times 3}$
$b \xrightarrow{\times 2}$ → $\times$ → $3a(2b + c)$
c → $+$

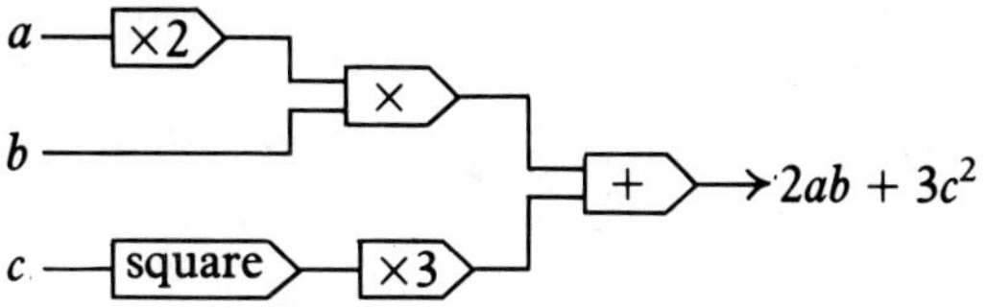

(f)

a
b → $\times$ → $\times$ → $\times 3$ → $3abc$
c

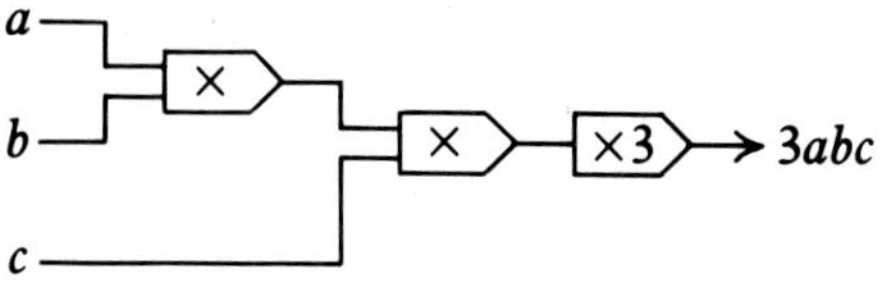

E Division

E1 (a) 5 (b) 7 (c) 5 (d) $\frac{1}{2}$

E2 (a) 2 (b) 10 (c) 5

E3 (a) 60 (b) 90 (c) 108 (d) 120
(e) 135 (f) 140 (g) 162

E4 (a) 2 (b) 3

E5 (a) 6 mg (b) 24 mg

E6 The load is 3072 kg.

E7 (a)

Speed v, in m.p.h.	0	10
Stopping distance s, in metres	0	5

20	30	40	50	60	70
13·3	25	40	58·3	80	105

(b) See graph. (c) About 26 m.p.h.

E8 The distance is 506 mm.

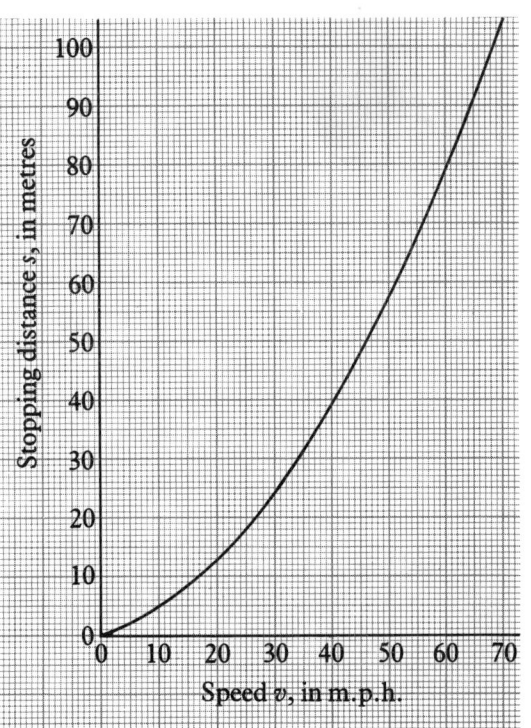

---

8 TV programmes survey

This is intended as a class activity. The class may be split up into groups to work on different channels. What will become evident – and this is an important part of the activity – is that people will differ about the classification of individual programmes, and there is no universally agreed 'right answer' for the percentage breakdown.

---

9 Standard index form (1)

In the early part of this chapter we show how large numbers can be expressed using 'units' such as millions. Standard index form is then introduced through the notion of order of magnitude. A number in standard form, e.g. $2·7 \times 10^8$, is thought of not as '2·7 multiplied by ten 8 times' but as 2·7 lots of 10^8, where the order of magnitude, 10^8, functions as a kind of 'unit'.

A Millions

A1 (a) 3 000 000 000, 400 000 000,
254 000 000, 120 000 000,
70 000 000, 29 000 000

(b) 2000 million, 532 million, 70 million,
60 million, 19 million, 8 million

(c) Most people will find the second set easier.

A2 (a) 34·5 million   (b) 13·9 million
    (c) 8·75 million   (d) 10·67 million
    (e) 4·76 million   (f) 0·36 million

A3 (a) 85·5 thousand   (b) 13·7 thousand
    (c) 20·9 thousand

A4 (b) 1·3 million   (c) 2·3 million
    (d) 1·0 million   (e) 7·0 million

B   Powers of 10

B1 (a) 100 000   (b) 100 000 000   (c) 1 000 000

B2 (a) 10^3 (b) 10^2 (c) 10^1

B3 (a) 8×10^4 (b) 8×10^1
    (c) 8×10^5 (d) 8×10^6

B4 (a) $4 000 000 = 4 \times 10^6$
    (b) $600 000 000 = 6 \times 10^8$
    (c) $20 000 = 2 \times 10^4$
    (d) $50 000 000 000 = 5 \times 10^{10}$

B5 (a) 5×10^6 (b) 7×10^{13} (c) 6×10^4

B6 (a) 50 000 000   (b) 300
    (c) 400 000 000   (d) 90

B7 4×10^3, 9×10^4, 2×10^5,
    6×10^5, 7×10^8

B8 5×10^9, 6×10^9, 8×10^{13},
    8×10^{17}, 2×10^{23}

C   Standard index form

C1 $4·2 \times 10^5$

C2 (a) $9·3 \times 10^6$ (b) $2·7 \times 10^4$ (c) $5·6 \times 10^2$
    (d) $4·8 \times 10^7$ (e) $4·85 \times 10^7$

C3 (a) $4·2 \times 10^4$ (b) $7·5 \times 10^5$
    (c) $8·23 \times 10^6$ (d) $5·3 \times 10^2$
    (e) $4·07 \times 10^9$

C4 $6·6 \times 10^{22}$ metres

C5 (a) 7300   (b) 73 000   (c) 73 000 000
    (d) 824 000   (e) 824   (f) 90 600
    (g) 1 080 000   (h) 472 400 000

C6 14 000 000 000 000 000 000 000 metres

C7 (a) α Canis Majoris (Sirius)
    (b) β Centauri (Hadar)

C8 (a) Mercury
    (b) Mercury, Pluto, Mars, Venus, Earth,
       Neptune, Uranus, Saturn, Jupiter
    (c) Mercury, Pluto, Mars, Venus, Earth,
       Uranus, Neptune, Saturn, Jupiter
    (d) Saturn   (e) 100 times

D   Scientific notation on a calculator

D1 The answer is given in the question.

D2 (a) 4×10^3 (b) $9·88 \times 10^{17}$
    (c) $1·3 \times 10^4$ (d) $9·0 \times 10^5$

D3 (a) $1·799 \times 10^{10}$ m   (b) $1·079 \times 10^{12}$ m
    (c) $2·590 \times 10^{13}$ m   (d) $9·461 \times 10^{15}$ m
    (all to 4 s.f.)

D4 (a) $5·7 \times 10^{19}$ m   (b) $1·3 \times 10^{21}$ m
    (c) $2·1 \times 10^{22}$ m   (all to 2 s.f.)

D5 About $2·2 \times 10^9$ times

10 Percentage (1)

This chapter covers finding a percentage of a quantity and percentage increases and decreases. The 'multiplier' method is used for percentage changes (e.g. a 35% increase means multiply by 1·35) as it is a more powerful method in general. (It is the one used in the Y books, which some pupils may move on to later.) However, if this method is found to be too difficult, then the more straightforward method (of calculating 35%, say, and adding it on) may be used.

A Calculating a percentage of a quantity

A1 382·5 g of fat

A2 (a) 0·3   (b) 75 g of fat

A3 (a) 0·2   (b) 170 g

A4 (a) 0·14   (b) 0·04   (c) 0·09
(d) 0·8   (e) 0·03

A5 (a) 73%   (b) 70%   (c) 8%   (d) 41%
(e) 6%

A6 (a) 1·65 g   (b) 3·6 g   (c) 3·15 g

A7 (a) 2·45 g   (b) 1·8 g

A8 55 g of fruit and nut cereal contains
6·05 g of protein.
30 g of steak and kidney casserole
contains 7·2 g of protein.
So steak and kidney casserole has more
protein.

A9 (a) $2·2 + 7·2 + 3·15 = 12·55$ g
(b) $1·4 + 1·8 + 1·65 = 4·85$ g
(c) $15·2 + 18·0 + 8·7 = 41·9$ g

B Calculating percentages in your head

B1 (a) £20   (b) £13   (c) 20p   (d) 8p
(e) £6   (f) 9p   (g) £24   (h) £360

B2 (a) 38p   (b) £1·26   (c) £1·30
(d) £10·38   (e) 0·58 m   (f) 0·396 kg

B3 (a) £45   (b) £40·50   (c) £16·65
(d) £6·75

C Using your common sense

C1 (a) £22·80   (b) £65·10   (c) £5·04
(d) £5·76   (e) £116·80   (f) £27·06

C2 (a) Miss Murphy gets £176,
Mr Patel gets £144.
(b) Miss Murphy gets £363,
Mr Patel gets £297.
(c) Miss Murphy pays £69·30,
Mr Patel pays £56·70.

D Percentage decreases

D1 (a) £7·80   (b) £28·60   (c) £2·34

D2 (a) 55%   (b) 0·55   (c) £10·12

D3 (a) 88%   (b) £198

D4 (a) 0·85   (b) £41 310

D5 (a) 0·74   (b) £281·20

D6 (a) 34·32 kg   (b) £242·88
(c) 71·76 km   (d) £526·71

E Percentages greater than 100%

E1 (a) 1·4   (b) 10·5 kg

E2 (a) £87·10   (b) 38·4 m   (c) 133·2 m

E3 (a) 1·6   (b) 1·63   (c) 1·06
(d) 1·39   (e) 1·9

F Percentage increases

F1 124 500

F2 (a) 97·2 million   (b) 58·8 million
(c) 29·7 million

F3 £643·50

F4 (a) 1·45   (b) £40·60

F5 (a) 108·54 kg   (b) £115·62
(c) 36·48 kg   (d) £45·50

F6 (a) 1·07   (b) £69·55

F7 (a) £94·76   (b) £376·48
(c) £98·28   (d) 59·16 m

F8 Alfred's new salary is £10 272·40.
Alberta's new salary is £10 189·92.
So Alfred earns more afterwards.

11 Negative numbers

Addition and subtraction of negative numbers appeared in *Book B1*. This chapter starts with a second look at these operations, approached through number patterns. A number pattern approach is also used for multiplication and division.

A Negative numbers: addition and subtraction

A1 (a) $^-1$ (b) 2 (c) $^-3$ (d) $^-4$ (e) $^-4$
(f) $^-5$

A2 $4 + {}^-6 = {}^-2$
$4 + {}^-7 = {}^-3$

A3 (a) $5 + {}^-3$ (b) $2 + {}^-6$
　　$= 5 - 3$ 　　$= 2 - 6$
　　$= 2$ 　　　$= {}^-4$

(c) $^-3 + {}^-4$
　$= {}^-3 - 4$
　$= {}^-7$

A4 (a) 3 (b) $^-3$ (c) $^-5$ (d) $^-7$ (e) $^-4$
(f) 0 (g) $^-7$ (h) $^-2$

A5 $7 - {}^-5 = 12$
$7 - {}^-6 = 13$

A6 (a) 8 (b) 10 (c) 12

A7 (a) 7 (b) $6 - {}^-1 = 7$
(c) 8 (d) $6 - {}^-2 = 8$

A8 (a) $^-1 - {}^-3$ (b) $3 - {}^-4$
　　$= {}^-1 + 3$ 　　$= 3 + 4$
　　$= 2$ 　　　$= 7$

(c) $^-8 - {}^-6$
　$= {}^-8 + 6$
　$= {}^-2$

A9 (a) 7 (b) 5 (c) 3 (d) 2 (e) 4
(f) 4 (g) 10 (h) $^-7$ (i) 1 (j) 6
(k) 8

A10 (a) (i) 30 (ii) 50 (iii) 10
(b) $^-10$. The inside is colder than the outside.
(c) $^-30$

A11 (a) 4 (b) $^-4$ (c) $^-5$ (d) 5 (e) $^-4$
(f) $^-7$ (g) 10 (h) $^-6$ (i) $^-5$ (j) 8
(k) 2

A12 (a) $^-7$ (b) 0 (c) 12 (d) $^-1$
(e) $^-11$ (f) 6 (g) 6 (h) 13 (i) $^-7$
(j) 3 (k) $^-5$

A13 (a) $^-7$ (b) $^-3$ (c) $^-3$ (d) $^-4$ (e) 0
(f) 0 (g) $^-11$ (h) $^-2$ (i) $^-15$ (j) 8
(k) $^-5$

A14 (a) $4 + {}^-5 - {}^-2 = 1$
(b) $4 + {}^-2 - {}^-5 = 7$
(c) $^-5 + {}^-2 - 4 = {}^-11$

A15 (a) $^-4 - 3 - {}^-3 = {}^-4$
(b) $^-3 - 3 - {}^-4 = {}^-2$
(c) $3 - {}^-4 - {}^-3 = 10$

A16 $3 - \boxed{{}^-2} + {}^-5 = 0$

A17 $5 - \boxed{{}^-11} - {}^-4 = 20$

B Multiplication

B1 (a) $^-15$ (b) $^-40$ (c) $^-12$ (d) $^-56$
(e) $^-27$ (f) $^-16$ (g) 0 (h) $^-42$

B2 (a) $^-10$ (b) $^-12$ (c) $^-14$ (d) 0
(e) $^-54$ (f) $^-32$ (g) $^-24$ (h) $^-49$

B3 $c = 13$

B4 $z = {}^-22$

B5 $w = {}^-5$

B6 $c = {}^-31$

B7 $k = 4$

B8 $t = {}^-9$

B9 (a) 16 (b) 12 (c) 20 (d) 21
(e) 14 (f) 24 (g) 35 (h) 4
(i) 64 (j) 36 (k) 36 (l) 36

B10 (a) $^-35$ (b) 16 (c) $^-27$ (d) $^-42$
(e) 32 (f) $^-24$ (g) 25 (h) 9 (i) 1
(j) 0

C Division

C1 (a) 5 (b) ⁻6 (c) ⁻10 (d) 4 (e) ⁻2
(f) ⁻7 (g) 4 (h) 0 (i) ⁻9 (j) ⁻8½
(k) ⁻15½

C2 (a) 45 (b) ⁻11 (c) 8 (d) 0 (e) 20
(f) 0 (g) ⁻15 (h) 9 (i) ⁻1 (j) ⁻48
(k) 49

C3 (a) 4 (b) ⁻12 (c) ⁻5 (d) ⁻6
(e) 12 (f) 8 (g) 9 (h) ⁻15 (i) ⁻1
(j) ⁻1 (k) ⁻6 (l) 5 (m) 36 (n) 5
(o) 16

C4 $c = 30$

C5 $z = {}^{-}4$

C6 $s = 9$

C7 $k = {}^{-}14$

C8 $c = 6$

C9 $n = {}^{-}2$

C10 $r = 13$

D Negative numbers on a calculator

D1 Should be checked by pupil's calculator

D2 (a) ⁻2·1 (b) 1·3 (c) ⁻4·5 (d) ⁻7
(e) ⁻2·7 (f) 3·24 (g) ⁻3·4 (h) 6·5

D3 $s = 3\cdot096$

Review 2

---

7 The language of algebra

7.1 (a) 23 (b) 17 (c) 35 (d) 28
(e) 41 (f) 52 (g) 5 (h) 18

7.2 (a) C (b) B (c) A

7.3 (a) 20 (b) 8 (c) 64 (d) 144
(e) 12

7.4 (a) 12 (b) 9 (c) 4 (d) 108
(e) 192 (f) 8 (g) 4 (h) 4½
(i) 288 (j) 6

7.5 (a) $a(b + c) = 30$ (b) $(ab)^2 = 400$
(c) $a(bc)^2 = 320$ (d) $a(bc + 3c) = 70$
(e) $\frac{1}{4}(ac)^2 = 25$ (f) $(ab + c)^2 = 484$

7.6 (a) $3p + q^2$ (b) $3(p + 4q)^2$

7.7 (a) 10 (b) 1 (c) 7 (d) 4 (e) 3·6

7.8 (a) 1·76 (b) 50·18 (c) 164·84

9 Standard index form (1)

9.1 (a) 4·5 million (b) 13·6 million
(c) 23·5 million (d) 3·13 million

9.2 (a) $5\cdot23 \times 10^6$ (b) $6\cdot75 \times 10^3$
(c) $1\cdot037 \times 10^4$ (d) $9\cdot3 \times 10^{10}$

9.3 (a) 27 000 (b) 1 080 000
(c) 372 000 (d) 41 600 000

9.4 (a) α Centauri (Rigil Kent)

(b) β Orionis (Rigel)
(c) α Aurigae (Capella)
(d) α Eridani (Achernar)

9.5

Star	Distance in light-years (to 3 s.f.)
α Centauri (Rigil Kent)	4·33
α Lyrae (Vega)	26·5
α Boötes (Arcturus)	35·9
α Aurigae (Capella)	44·4
β Orionis (Rigel)	1160
α Eridani (Achernar)	137

9.6 1410 kg/m³ (to 3 s.f.)

10 Percentage (1)

10.1 (a) 0·32 (b) £27·20

10.2 (a) 0·56 (b) 0·04 (c) 0·20
(d) 0·99 (e) 0·01

10.3 (a) 170·2 kg (b) 2·72 m (c) £364

10.4 (a) 1·3 (b) 1·1 (c) 1·46
(d) 1·05 (e) 1·83

10.5 (a) 1·38 (b) £62·10

10.6 (a) £61·48 (b) £87·74

10.7 (a) 88% (b) 0·88 (c) £49·28

10.8 (a) 0·66 (b) £211·20

10.9 (a) £15·47 (b) 201·6kg

11 Negative numbers

11.1 (a) ⁻15 (b) ⁻1 (c) 4 (d) 2 (e) 9
  (f) ⁻9 (g) ⁻6 (h) ⁻4 (i) ⁻16
  (j) ⁻3

11.2 (a) 5 (b) 5 (c) ⁻5 (d) 11
  (e) ⁻9 (f) ⁻3 (g) ⁻4 (h) 6 (i) 0
  (j) ⁻3

11.3 (a) ⁻2 (b) ⁻8 (c) 5 (d) ⁻2 (e) 7
  (f) 5 (g) 3 (h) ⁻10 (i) ⁻13 (j) 2

11.4 (a) 5 (b) ⁻20 (c) 20 (d) ⁻6 (e) ⁻18
  (f) ⁻1 (g) 0 (h) 2 (i) 0 (j) 7

11.5 (a) $c = 12$ (b) $c = 0$
  (c) $c = {}^-19$ (d) $c = 19$

11.6 (a) $s = {}^-6$ (b) $s = {}^-20$

11.7 (a) $z = {}^-3$ (b) $z = {}^-6$

M Miscellaneous

M1 The final entries in the table will be:

		too small	too big
Number	. . .	2·71	2·72
Cube	. . .	19·9025	20·1236

12 Loci

Sections A and B contain a number of concrete examples which illustrate the general idea of a locus, introduced in section C. The loci throughout the chapter are *regions*.

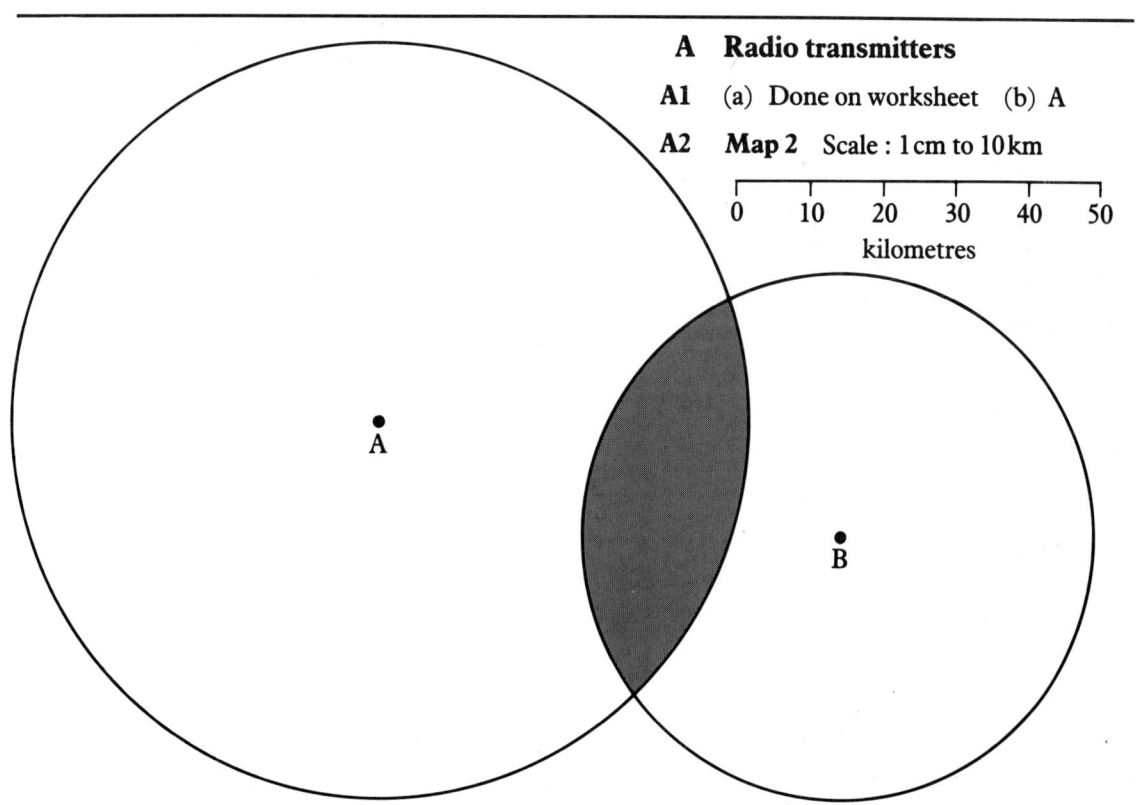

A Radio transmitters

A1 (a) Done on worksheet (b) A

A2 **Map 2** Scale : 1 cm to 10 km

0   10   20   30   40   50
kilometres

A3 The major reason is that the land is not flat. Reception will be heavily dependent on the height of the land.

B Mainly gardens

The diagrams drawn here are for identification purposes. They are usually reduced in scale compared with the actual size.

B1

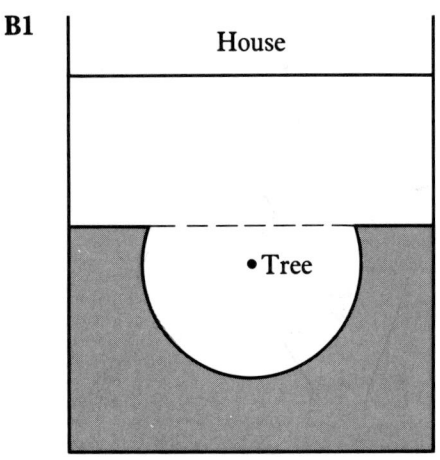

B2

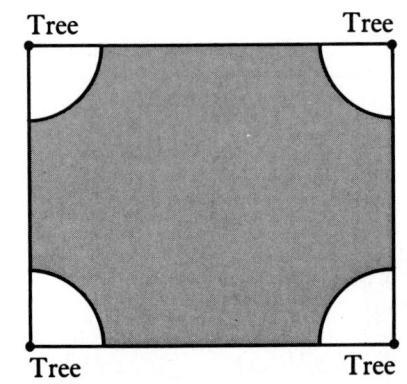

B3

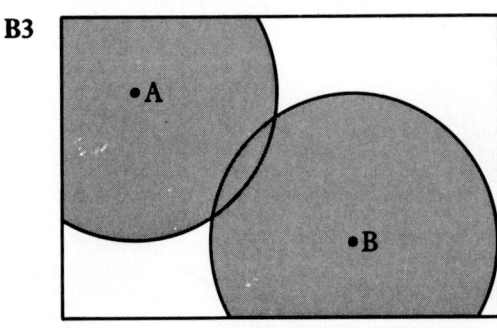

B4

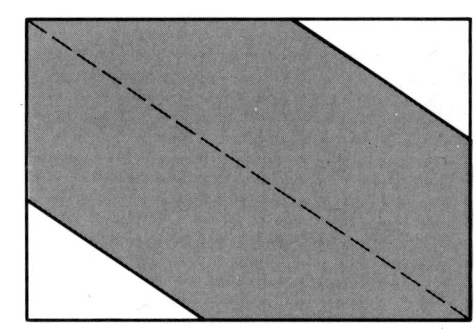

B5

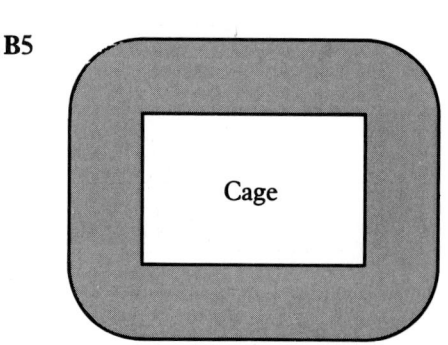

B6 RURITANIA

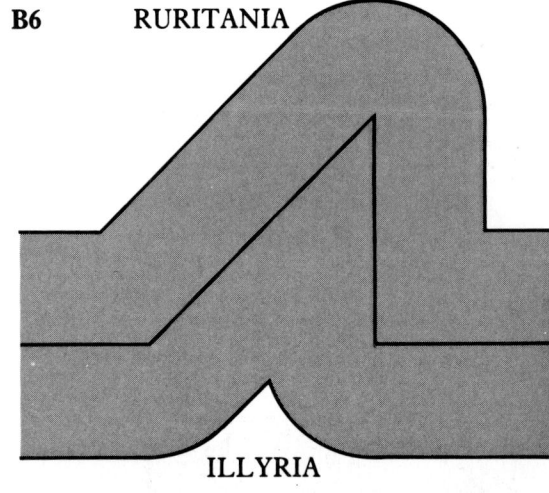

ILLYRIA

C Loci

C1

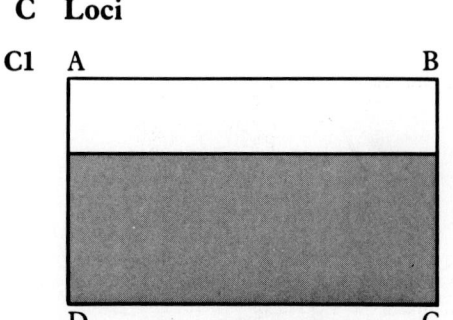

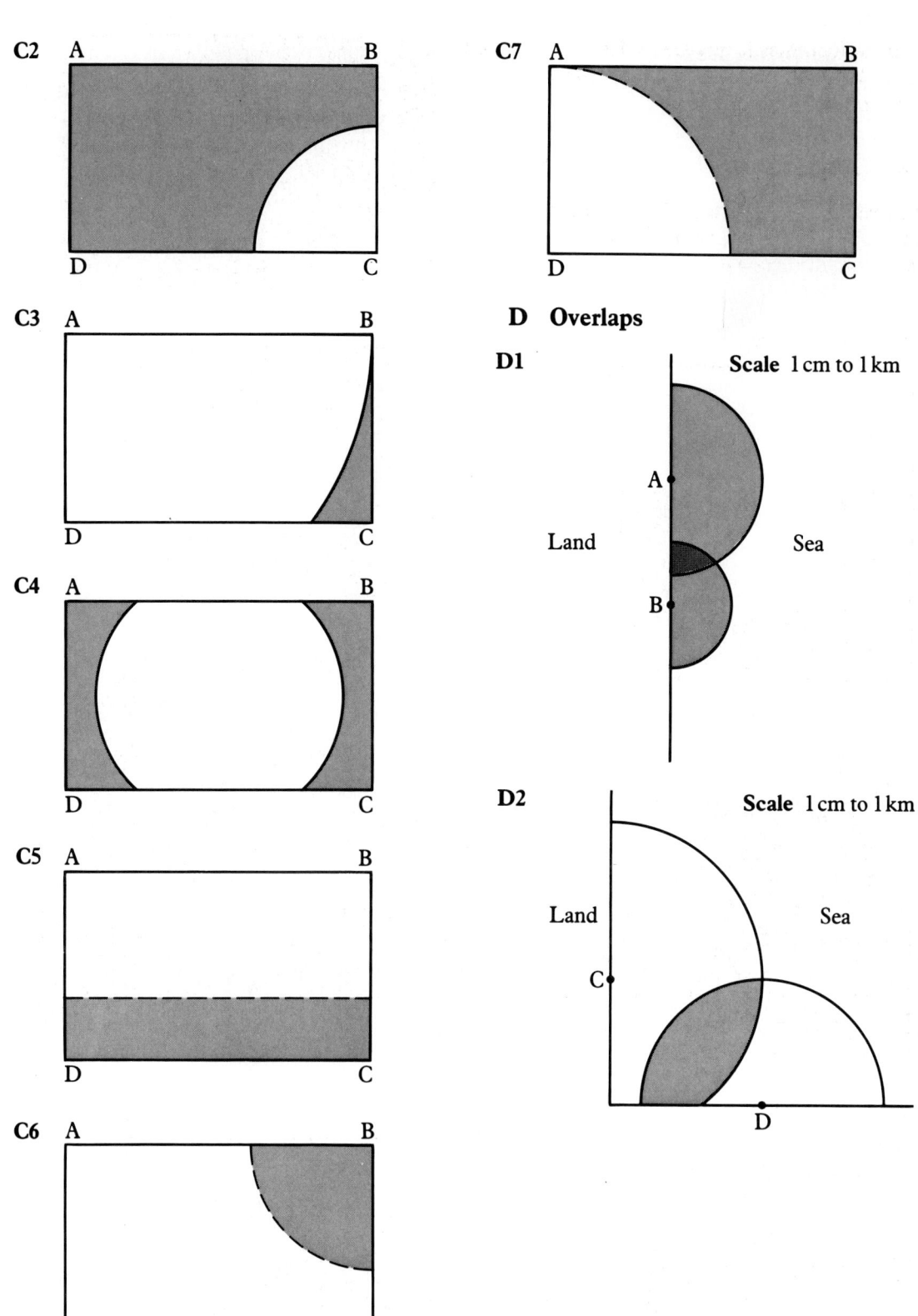

C2

A　　　　　　　　　B

D　　　　　　　　　C

C3

A　　　　　　　　　B

D　　　　　　　　　C

C4

A　　　　　　　　　B

D　　　　　　　　　C

C5

A　　　　　　　　　B

D　　　　　　　　　C

C6

A　　　　　　　　　B

D　　　　　　　　　C

C7

A　　　　　　　　　B

D　　　　　　　　　C

D Overlaps

D1   **Scale** 1 cm to 1 km

A

Land   Sea

B

D2   **Scale** 1 cm to 1 km

Land   Sea

C

D

E Seeing and being seen

E1 (a) She can only be seen from P and R.
(b)

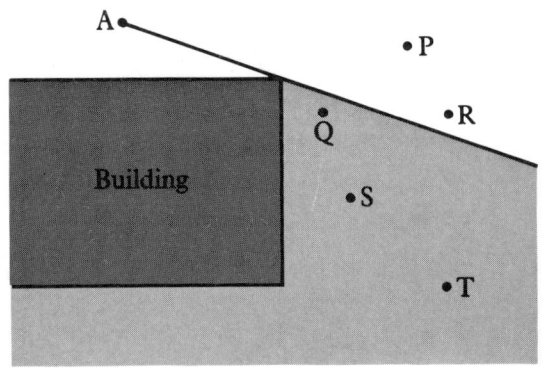

(c)

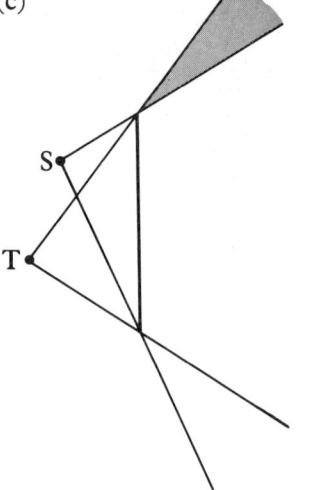

E2

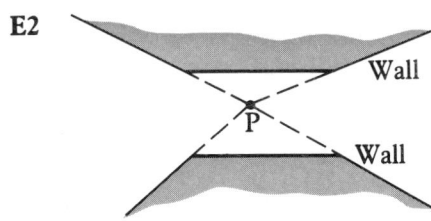

(d)

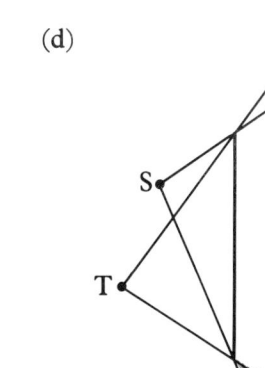

E3 (a)

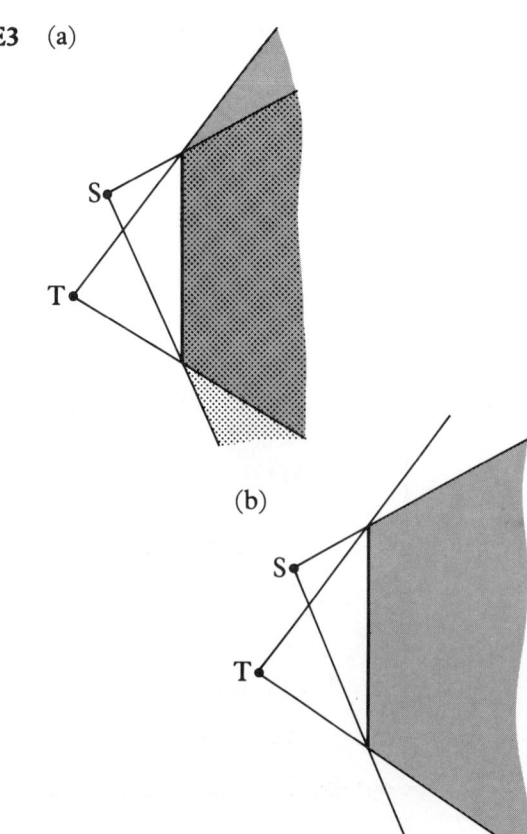

(b)

(e)

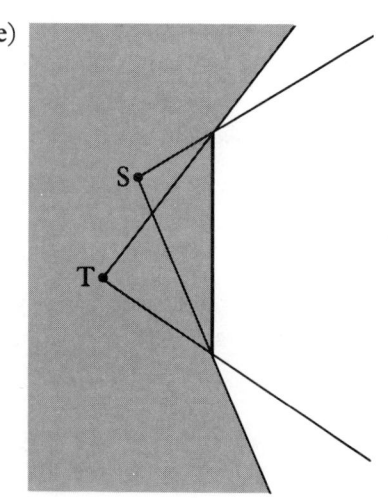

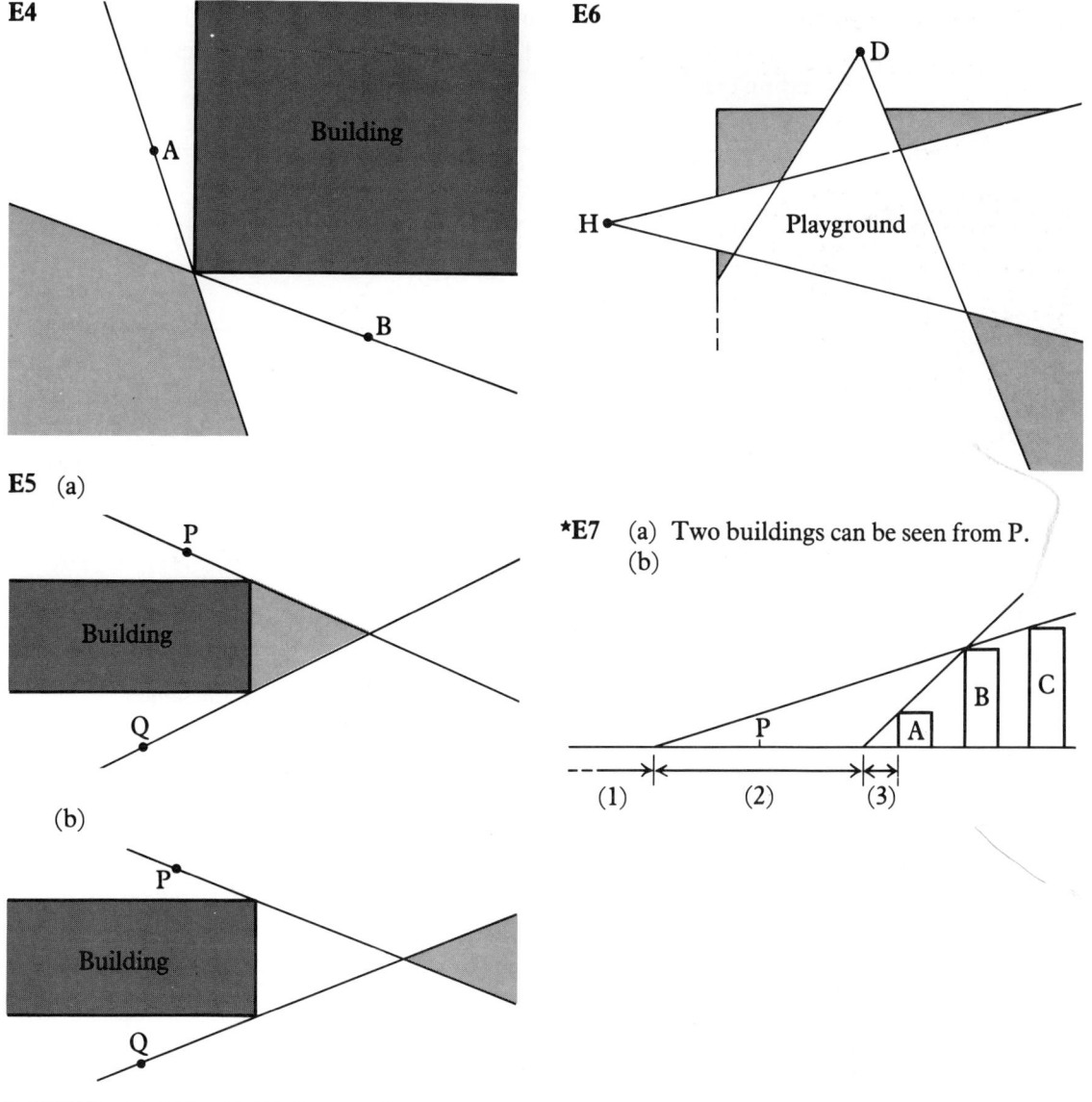

E4

Building

A

B

E6

D

H

Playground

E5 (a)

P

Building

Q

(b)

P

Building

Q

*E7 (a) Two buildings can be seen from P.
(b)

P

A

B

C

(1) (2) (3)

13 Ratio

This is a very important chapter. The ideas in it arise frequently in later
work. Throughout the series the notation used for a ratio is $\frac{a}{b}$, and a
ratio is expressed numerically as a single number. We are following the
usage whereby we say, for example, that 'in a circle the ratio
$\frac{\text{circumference}}{\text{diameter}}$ is 3·14 (to 3 s.f.)'. The ratio $\frac{a}{b}$ is identified
with the multiplier from b to a (i.e. the number by which you have to
multiply b to get a).

Note that pupils' answers may be slightly different from those given here and still be acceptable, the differences being caused by measurement rounding errors.

A Multipliers

A1 The multiplier is 4.

A2 (a) 5   (b) 8   (c) 4   (d) 4

A3 (a) 13   (b) 4·5   (c) 3·5   (d) 3·5   (e) 16
(f) 55   (g) 12·5

A4 You cannot compare a length with a weight in this way.

A5 (a) 3·86   (b) 4·21   (c) 1·79
(d) 4·72   (e) 12·49

A6 (a) (i) 24·50   (ii) 18·31   (iii) 16·67
(iv) 11·06
(b) Blagdon Park First School

A7 (a) Polar bear 800,   black bear 369,
chimpanzee 36,   kangaroo 3750,
tortoise 133
(b) Kangaroo   (c) Chimpanzee

A8 $\dfrac{11}{6}$ **A9** (a) $\dfrac{4}{7}$ (b) $\dfrac{6}{5}$

A10 (a) 0·750   (b) 1·44   (c) 0·915   (d) 0·663

A11 Melissa is taller.   **A12** p is larger.

B 'No change' rule

B1 (a) 1·6   (b) 1·6   (c) 1·6

B2 (a) (i) 1·25   (ii) 0·625   (iii) 0·5
(b) (i) 0·625   (ii) 1·25   (iii) 1·25   (iv) 0·5

B3 (a) 2·24   (b) 2·24

C Enlargement

C1 (a) 3·0 cm   (b) 4·8 cm   (c) 4·8 cm

C2 (a) 5·6 cm   (b) 8·96 cm
(c) 9·0 cm (to 1 d.p.)

C3 Approximately 1·67

C4 (a) 55 mm   (b) 3·3   (c) 129 mm

C5 0·8 (to 1 d.p.)

C6 (a) 0·8   (b) 0·8

D The effect of enlargement on the ratio of two lengths

D1

Scale factor of enlargement	1	2	3	4
Height of tower, in mm	28	56	84	112
Length of church, in mm	35	70	105	140
Ratio $\dfrac{\text{height}}{\text{length}}$	0·8	0·8	0·8	0·8

D2 (a) 140 mm   (b) 175 mm   (c) 0·8

D3 0·8

D4 (a) 1·4   (b) The ratio is the same.

D5 (a) Knife P: blade 22 mm, handle 18 mm
Knife Q: blade 44 mm, handle 35 mm
(b) 1·22   (c) 1·26
(d) It tells you the scale factor.
(The difference between (b) and (c) is due to a measurement rounding error.)

D6 (a) $\dfrac{\text{length of Y}}{\text{length of X}} = \dfrac{\text{height of Y}}{\text{height of X}} = 1\cdot3$
(b) $\dfrac{\text{length of Y}}{\text{height of Y}}$

E Similarity

E1 (a) A 48 mm, B 72 mm
(b) Scale factor 1·5

E2 (a) A 40 mm, B 60 mm
(b) Scale factor 1·5

E3 The ratio for both shapes is 1·2.

E4 (a) 1·5   (b) It could fit rectangles C and D.

E5 (a) Between 1·41 and 1·42
(b) Between 1·41 and 1·42
(c) Between 1·41 and 1·42
All the rectangles are similar.

14 Pythagoras' rule

Section A revises the work on Pythagoras' rule done in chapter 6.
Sections B and C extend the work to calculating one of the shorter sides
of a right-angled triangle.

A Calculating the longest side of a right-angled triangle

A1 (a) 4·6cm   (b) 3·1cm   (c) 4·5cm

B Calculating one of the shorter sides

B1 (a) 5·29   (b) 4·36   (c) 4·90

B2 (a) 6·00   (b) 4·47   (c) 7·55
    (d) 6·24   (e) 7·21   (f) 9·17   (g) 9·38

C Mixed problems

C1 5·7m

C2 12·0cm

C3 1·3m

C4 3·5m

C5 (a) 30·0cm   (b) 45·8cm

15 Percentage (2)

This chapter includes work on expressing one quantity as a percentage of
another, and expressing an increase or decrease as a percentage increase
or decrease. The multiplier method for expressing increases and
decreases as percentages is recognised as being difficult and this work is
starred.

A Expressing one amount as a percentage of another

A1 15%

A2 62·5%

A3 8%

A4 52%

A5 (a) 87%   (b) 4%   (c) 42%   (d) 67%

    19%

A7 18%

A8 29%

A9 19%

A10 52% boys,   48% girls

A11 (a) 59   (b) 36%   (c) 59%

    (d) 57%   (e) 118   (f) 39%
    (g) 15%   (h) 242   (i) 24%

A12 34%

B Drawing pie charts

B1 (a) 30%   (b) 14%

B2

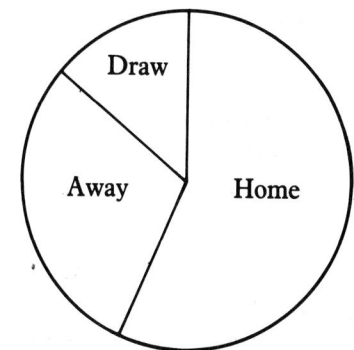

B3 (a) 54·5% home, 18·2% away,
27·3% draw
(b)

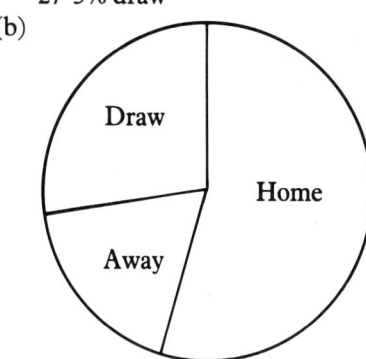

B4 (a)

1979	
Conservative	44·9%
Labour	37·7%
Liberal	14·1%
Others	3·3%

1983	
Conservative	43·5%
Labour	28·3%
Liberal/SDP	26·0%
Others	2·2%

(b)

Others

Liberal

Conservative

Labour

1979

Others

Liberal/
SDP

Conservative

Labour

1983

C Percentage decreases and increases : review

C1 (a) 78% (b) £7·41

C2 (a) £128·80 (b) £54·60
(c) £77·83 (d) £125·56

C3 (a) 1·13 (b) £6474·90

C4 (a) £206·50 (b) £130·68
(c) £43·68 (d) £277·86

D Expressing a decrease as a percentage decrease

D1 22%

D2 (a) £13 (b) 19%

D3 (a) 19% (b) 8% (c) 11% (d) 14%

D4 28% (to nearest 1%)

★D5 (a) 26% (b) 10% (c) 54%
(d) 27% (e) 12% (f) 13%

★D6 23%

★D7 20%

E Expressing an increase as a percentage increase

E1 7%

E2 15%

E3 (a) 33% (b) 42% (c) 4% (d) 16%

E4 (a) 23% decrease (b) 29% increase
(c) 8% increase (d) 12% decrease

★E5 7%

★E6 8%

★E7 (a) 19% (b) 33% (c) 26%
(d) 6% (e) 7% (f) 14%

★E8 10%

★E9 (a) 30·1 m.p.g. (b) 32·3 m.p.g.
(c) 7%

★E10 (a) 58·8p (b) 18% (to nearest 1%)

F Miscellaneous questions on percentage

All percentage answers here are given to the nearest 1%.

F1 68%

F2 594 cycles

F3 (a) 50% (b) 37% (c) £7440·80 (d) £7069·80 (e) 5% decrease

F4 (a) 68% (b) 78% (c) 73% (d) 10% (e) 33% (f) 4%

F5 (a) 30% (b) 28% (c) 29% (d) 54% (e) 46% (f) 48% (g) 16%

F6 (a) 34% (b) Pagan 50%, Jovan 50%

F7 The Baxi 511 RS wall mounted boiler

F8 Reduce the amount of chips by 20%.

---

16 Investigations (2)

These investigations, based on the Morse code, are intended as a class activity. Although pupils may work alone on parts of the investigation, it is intended that there should be discussion of methods and results as the work proceeds. There are some further 'Morse code' investigations in the review section, which may be used as individual follow-up work.

A The Morse code

A1 DO
NOT
DECODE
THIS
MESSAGE

A2 There are only 30 possible codes without extending to five symbols. The most efficient codings of the Russian alphabet will therefore have just one letter with a five-symbol code.

A3 He considered E and T to be the most commonly used letters.

A4 (a) 4 (b) 8 (c) 16

A5 The number of extra letters doubles each time because each of the possibilities for the previous stage can be extended with either a dash or a dot; for example, the 4 possible two-symbol codes

• •   • —   — •   — —

give rise to the following 8 three-symbol codes

• • —   • — —   — • —   — — —
• • •   • — •   — • •   — — •

B How many sequences?

B1 (a) 1 (b) 2 (c) 3 (d) 4 (e) 5

B2

Number of dashes	0	1	2	3	4	5	6	7
Number of different sequences with 1 dot	1	2	3	4	5	6	7	8

B3 (a) 1 (b) 3 (c) 6 (d) 10

B4

		Number of dashes					
		0	1	2	3	4	5
	1	1	2	3	4	5	6
Number of dots	2	1	3	6	10	15	21
	3	1	4	10	20	35	56
	4	1	5	15	35	70	126
	5	1	6	21	56	126	252

Lots of number patterns are present in the table, which is essentially a form of Pascal's triangle.
Each number is the sum of the number above it and the number on the left.
[Some pupils might be asked to explain why this should be so.]

17 Standard index form (2)

This chapter deals with numbers less than 1.

A Large numbers : review

A1 (a) 10^5 (b) 4×10^5
(c) $4 \cdot 2 \times 10^5$ (d) $42\,000\,000$

A2 (a) 6×10^4 (b) $6 \cdot 3 \times 10^4$
(c) $6 \cdot 37 \times 10^4$ (d) $4 \cdot 8 \times 10^6$
(e) $3 \cdot 5 \times 10^8$ (f) $2 \cdot 83 \times 10^{10}$
(g) $6 \cdot 43 \times 10^5$

A3 (a) 2900 (b) $8\,040\,000$
(c) $716\,000\,000$ (d) $34\,000$

B Negative powers of 10

B1 (a) 8×10^{-3} (b) 8×10^{-1}
(c) 8×10^{-4} (d) 8×10^{-7}

B2 (a) $0 \cdot 003 = 3 \times 10^{-3}$
(b) $0 \cdot 00004 = 4 \times 10^{-5}$
(c) $0 \cdot 09 = 9 \times 10^{-2}$
(d) $0 \cdot 00000005 = 5 \times 10^{-8}$

B3 (a) $0 \cdot 006$ (b) $0 \cdot 5$ (c) $0 \cdot 00003$
(d) $0 \cdot 08$ (e) $0 \cdot 000002$ (f) $0 \cdot 0004$
(g) $0 \cdot 00000009$ (h) $0 \cdot 0000003$

B4 3×10^{-7}, 8×10^{-5}, 6×10^{-4},
9×10^{-4}, 6×10^{-1}

B5 3×10^{-4}, 3×10^{-3}, 8×10^{-3}, 4×10^3,
7×10^4

B6 (a) $6 \cdot 5 \times 10^{-3}$ (b) $1 \cdot 3 \times 10^{-5}$
(c) $4 \cdot 38 \times 10^{-2}$ (d) $6 \cdot 275 \times 10^{-3}$

B7 (a) $3 \cdot 7 \times 10^{-3}$ (b) $9 \cdot 3 \times 10^{-13}$

(c) $8 \cdot 3 \times 10^{-2}$ (d) $1 \cdot 86 \times 10^{-4}$
(e) $6 \cdot 4 \times 10^{-5}$ (f) $5 \cdot 3 \times 10^{-7}$
(g) $9 \cdot 81 \times 10^{-6}$

B8 (a) $0 \cdot 00032$ (b) $0 \cdot 068$ (c) $0 \cdot 68$
(d) $0 \cdot 0000536$ (e) $0 \cdot 00402$
(f) $0 \cdot 0984$ (g) $0 \cdot 000000802$
(h) $0 \cdot 0003184$ (i) $0 \cdot 06249$

B9 (a) $4 \cdot 25 \times 10^4$ (b) $4 \cdot 25 \times 10^{-5}$
(c) $9 \cdot 44 \times 10^{-3}$ (d) $6 \cdot 13 \times 10^8$
(e) $7 \cdot 28 \times 10^5$ (f) $3 \cdot 02 \times 10^{-3}$
(g) $2 \cdot 93 \times 10^7$ (h) $7 \cdot 63 \times 10^{-5}$

B10 (a) $63\,000$ (b) $0 \cdot 00063$
(c) $0 \cdot 00000815$ (d) $724\,000$
(e) $3\,084\,000$ (f) $0 \cdot 0166$
(g) $0 \cdot 000083$ (h) 7100

C Using a calculator

C1 (a) 6×10^{-8} (b) 2×10^{-14}
(c) $6 \cdot 3 \times 10^{-3}$ (d) $5 \cdot 44 \times 10^5$
(e) $2 \cdot 3125 \times 10^9$ (f) $1 \cdot 344 \times 10^{-12}$
(g) $2 \cdot 624 \times 10^4$

C2

Country	Population density in persons per km^2 (to 3 s.f.)
UK	232
Holland	425
Belgium	325

Holland appears to be the most densely populated.

C3 The mass of one copper atom is $1 \cdot 06 \times 10^{-25}$ kg (to 3 s.f.).

Review 3

12 Loci

12.1

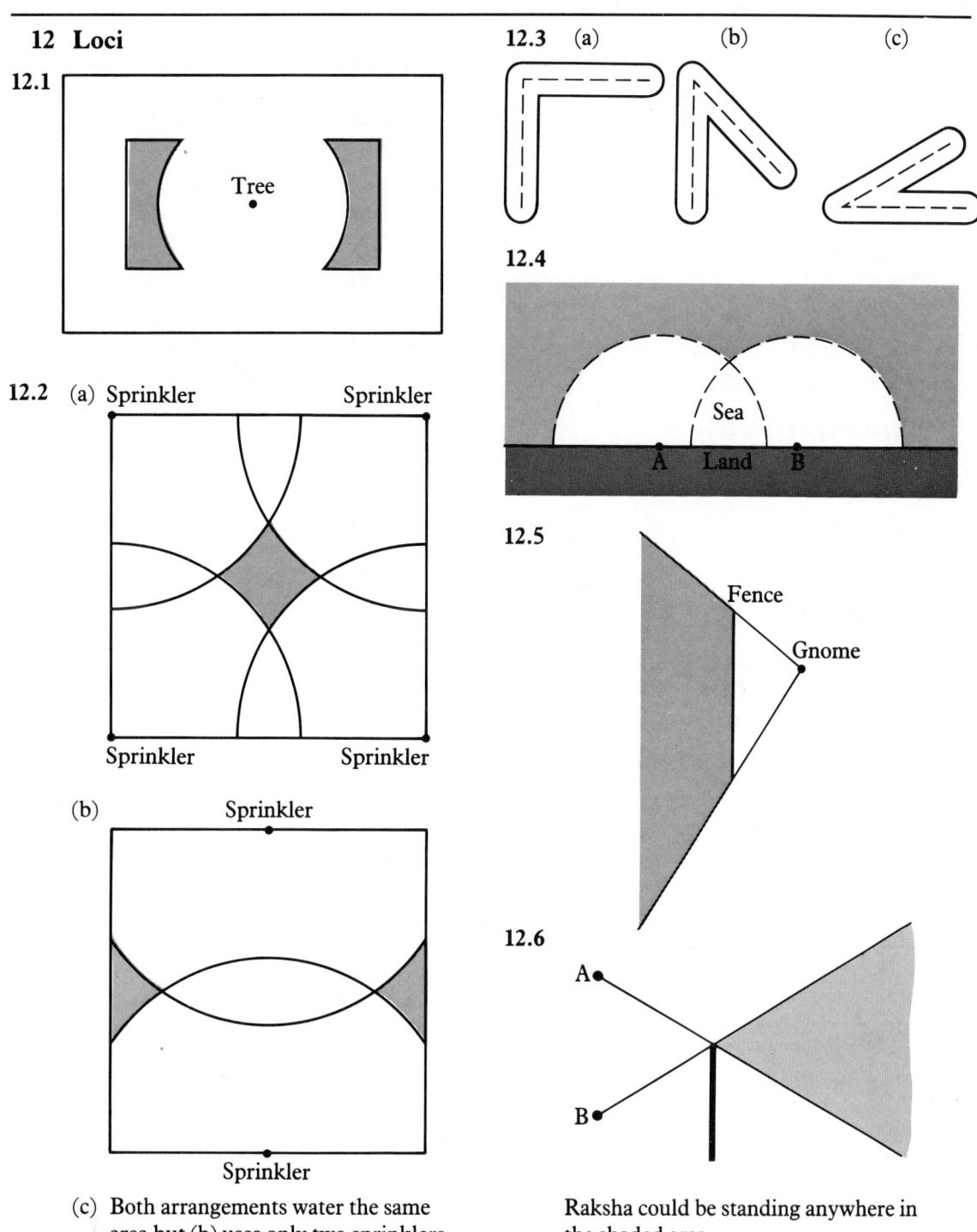

12.2 (a) Sprinkler   Sprinkler

Sprinkler   Sprinkler

(b)   Sprinkler

Sprinkler

(c) Both arrangements water the same area but (b) uses only two sprinklers.

12.3 (a)   (b)   (c)

12.4

Sea

A   Land   B

12.5

Fence

Gnome

12.6

A•

B•

Raksha could be standing anywhere in the shaded area.

13 Ratio

13.1 $\dfrac{0\cdot5}{1\cdot2}$

13.2 (a) $0\cdot826$ (b) $1\cdot07$ (c) $0\cdot0585$
(d) $0\cdot844$

13.3 (a) $1\cdot6$ (b) $1\cdot6$ (c) $1\cdot6$

13.4 (a) A $45\,\text{mm}$, B $65\,\text{mm}$,
scale factor $1\cdot44$
(b) $50\,\text{mm}$, $72\,\text{mm}$, ratio $1\cdot44$

13.5 (a) $\dfrac{w}{c}$, $\dfrac{v}{b}$, $\dfrac{u}{a}$ (b) $\dfrac{w}{u}$

14 Pythagoras' rule

14.1 (a) $9\cdot4$ (b) $4\cdot4$ (c) $12\cdot4$ (d) $3\cdot9$

14.2 $9\,\text{m}$ **14.3** $36\,\text{cm}$

15 Percentage (2)

15.1 42%

15.2 (a) $5\,130\,000$ (b) $48\cdot1\%$ (c) $48\cdot5\%$
(d) Women generally live longer than men.

15.3 (a) 6% (b) 20% (c) 24% (d) 23%

15.4 (a) 31% (b) 38% (c) 18% (d) 7%

15.5 (a) $58\cdot6\%$ (b) $57\cdot8\%$ (c) $41\cdot4\%$
(d) $53\cdot9\%$ (e) $43\cdot6\%$ (f) $56\cdot2\%$
(g) $19\cdot1\%$

16 Investigations (2)

16.1

Length of sequence	1	2	3	4	5	6	7	8	9	10	. . .
Number of different sequences	1	2	3	5	8	13	21	34	55	89	. . .

Each term (from the third term onwards) is the sum of the previous two terms.

This pattern arises because all the sequences of length 4, say, can be constructed either by adding a dash on the end of each sequence of length 2, or by adding a dot on the end of each sequence of length 3. This pattern continues indefinitely.

16.2

Length of sequence	1	2	3	4	5	6	7	8	9	10	. . .
Number of different sequences	1	1	2	3	4	6	9	13	19	28	. . .

In this sequence each term (from the fourth term onwards) is the sum of the previous term and the term three places back.

This pattern arises because all the sequences of length 4, say, can be constructed either by adding a dash on the end of each sequence of length 1, or by adding a dot on the end of each sequence of length 3.

17 Standard index form (2)

17.1 (a) 10^{-2} (b) 10^{-4} (c) 10^{-1} (d) 10^{-3}

17.2 (a) 3×10^{-3} (b) 6×10^{-2}
(c) 7×10^{-4} (d) 4×10^{-6}

17.3 (a) $3\cdot62 \times 10^{-4}$ (b) $5\cdot17 \times 10^{-6}$
(c) $3\cdot18 \times 10^{-1}$ (d) $9\cdot273 \times 10^{-5}$

17.4 (a) $0\cdot00003$ (b) $0\cdot7$
(c) $0\cdot00016$ (d) $0\cdot0000092$

17.5 (a) $2\cdot57 \times 10^7$ (b) $4\cdot83 \times 10^{-2}$
(c) $6\cdot72 \times 10^5$ (d) $1\cdot08 \times 10^{-4}$

M Miscellaneous

M1 $4\cdot4^2 = 19\cdot36$, so $4\cdot4 < \sqrt{20}$
$4\cdot5^2 = 20\cdot25$, so $4\cdot5 > \sqrt{20}$
Hence $4\cdot4 < \sqrt{20} < 4\cdot5$

M2 $a = 36°$, $b = 72°$
$c = 144°$, $d = 36°$

M3 (a) Yes (b) Yes (c) No
(d) No   (e) Yes

M4 It can be done in 5 regions like this:

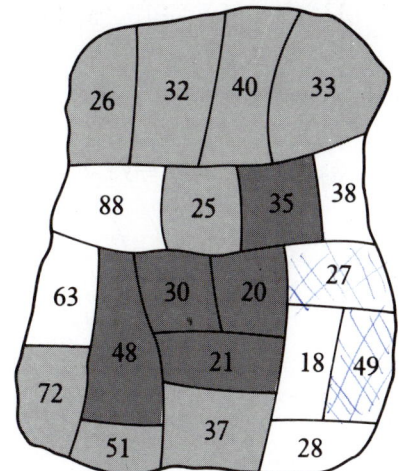